Hoodwinking the Nation

Julian L. Simon

Hoodwinking the Nation

Transaction Publishers
New Brunswick (U.S.A.) and London (U.K.)

A Cato Institute Book

First paperback printing 2006

This book is printed on acid-free paper that meets the American National Standard for Permanence of Paper for Printed Library Materials.

Library of Congress Catalog Number: 99-17327
ISBN: 1-56000-434-7 (cloth); 1-4128-0593-7 (paper)
Printed in the United States of America

Library of Congress Cataloging-in-Publication Data

Simon, Julian Lincoln, 1932-
Hoodwinking the nation / Julian L. Simon.
p. cm.
Includes bibliographical references and index.
ISBN 1-56000-434-7 (alk. paper)
1. Environmental sciences—Philosophy. 2. Environmental degradation. 3. Nature—Effect of human beings on. I. Title.

GE40.S55 ~~1999~~ 2006
363.7—dc21 99-17327
CIP

Contents

	Foreword	vii
	Preface	xiii
	Introduction	1
1.	What Do Americans Wrongly Believe about Environment, Resources, and Population?	7
2.	The Vanishing Farmland Scam	17
3.	The Concepts That Lead to Scares about Resources and Population Growth	31
4.	Why Does the Public Not Hear Sound Environmental Thinkers?	45
5.	Why Are So Many Biologists Alarmed?	55
6.	The Old-Time Journalistic Methods Don't Work Here	73
7.	Damn Lies, Statistics, and Doomsday	83
8.	Personal Knowledge versus Media-Shaped Opinions	93
9.	How Psychology Affects the Evaluation of Trends	101
10.	Why Do We Hear Prophecies of Doom from Every Side?	107
	Conclusion	123
	References	129
	Index	133

Foreword

Julian Simon wasn't wrong about many things, but his tragic and unexpected death at age 66 in February 1998 proved that, on one quite important matter, he was very wrong indeed.

Julian often felt, angrily at times, that he was being ignored, or ridiculed, by opponents in a vast Malthusian population-environment-resources conspiracy of crisis. He felt his work was not appreciated. There was some truth to the anti-Simon conspiracy idea. But insofar as a conspiracy existed, it existed precisely because Julian's work was not only appreciated by his opponents but feared for its intellectual potency and scholarly precision.

Somewhat ironically for a man who so valued measurement, Julian had little sense of where he stood on the scale of intellectual influence: Very, Very High. He would have been astonished at the encomiums that came to him in the days and weeks after his death. They came directly from the logical suspects, his supporters, and they were legion. And they came from elsewhere. The *New York Times* and the *Washington Post* published major serious essays about his fight against the conventional wisdom. Each placed him high in the firmament of what may be the central debate of our time: whether people are good for our planet or not. And in the *Boston Globe,* David Warsh, in writing of a 1980s "Washington School of intellectually powerful analysts determined to leverage their influence by proximity to power," said that Simon was "a big man, with big ideas, forever ready to explain."

Julian wanted to push a generation of Americans to rethink what they thought. By now it is clear that in this task he was largely successful. As the years roll on he will be more successful yet, his work studied, and picked at, by regiments of graduate students.

It is useful to remember some of the reasons why Julian succeeded.

His keystone work was *The Ultimate Resource*, published in 1981 and republished as *The Ultimate Resource 2* in late 1996. Its central point is clear: supplies of natural resources are not finite; they are created by the intellect of man, which is an infinitely renewable resource. (Coal, oil, and uranium were not resources until mixed well with intellect.)

The notion drove some enviros crazy. If it was true—poof!—there went so many of their crises. From air-conditioned offices accessed by computer-driven elevators, they brayed, "Simon believes in a technological fix!" The attacks were more than occasionally personal: after all, Simon's doctorate was merely in business economics; he was merely a professor of advertising and marketing; and, get this, he had actually started his own mail-order business and written a book about how to do it. (Never mind that he studied population economics for a quarter of a century and the mail-order book is still in print and in its fifth edition.)

But, irony again, it was Simon's knowledge of real-world commerce that gave him an edge in the intellectual wars. He knew first-hand about some things that many environmentalists of the time had only touched gingerly, like prices. If the ultimate resource was the human intellect, Simon reasoned, and the amount of human intellect was increasing both qualitatively and quantitatively, thanks to population growth, education, and technology, why, then, the supply of resources would grow, outrunning demand, pushing prices down, giving people more access to what they wanted, with more than enough left over to deal with pollution—in short, the very opposite of a crisis.

In all his work Simon rarely presented a sentence not referenced by facts, facts arranged in serried ranks to confront the opposition; facts about forests and food, pollution and poverty, nuclear power and nonrenewable resources; facts as foot-soldiers striking blows for accuracy.

In 1980 gloom-meister Paul Ehrlich accepted Simon's public offer of a wager. Malthusistically, Ehrlich bet that from 1980 to 1990 scarcity would drive resource prices up. Simonistically, Julian bet that abundance would push prices down. Simon won the bet. Ehrlich won a MacArthur "genius" award, which tells you more than you probably want to know about the MacArthur Foundation and geniuses. Not long before his death, Julian was ranked by *Washingtonian* magazine as one of the 25 smartest people in Washington. *Fortune* magazine listed Simon among the "150 Great Minds of the 1990s." Ehrlich didn't make the cut.

Simon sensed the primacy of something else that many environmentalists and crisis-mongers didn't catch on to for a quite a time. This: Human intellect could best be transformed into resources yielding beneficial goods and services in an atmosphere of political and economic liberty. I was a member of the Reagan-appointed U.S. delegation to the UN's Mexico City conference on population in 1984. Julian winced, and then counter-attacked, when population alarmists caricatured the American position as promoting the idea that "capitalism is the best contraceptive." It was not a good idea to ridicule capitalism, or free markets, or human liberty, in Julian's presence.

Well, as it turns out, rising living standards do tend to depress fertility. And living standards do rise faster under democratic market systems. And even academics now understand that the capitalist fruits of high economic growth can be used to diminish pollution. And there aren't a whole lot of serious folks still preaching that we're running out of everything. (Although a recent Worldwatch fundraising letter says the world may face danger because AIDS and water shortages will cut population.) One good memorial to Julian would be to put global warmists in the Simon-scope.

Most important is the kind of man Julian Simon was. Simon, the public personality, could sometimes glow like an exposed wire, crackling with nervous intellectual intensity. In the early 1980s I conducted a dialogue with Julian and Garrett Hardin, an environmentalist who had challenged Julian's facts and methods. Before the interview Julian was ashen faced. He asked for a private place to lie down. Then we sat down for the dialogue and Julian demolished Hardin. I would describe Julian's public personality as intense and competitive, but only because I can't think of tougher words. He could overcome severe hurdles, apparently by force of will, fueled by his sizzling energy.

But privately, he had a soul of purest honey, with something else going for him. He was a traditional man who believed that God created man to do the best he could, which was quite good indeed. Sabbath dinner at the Simon house was a gentle and joyous celebration. Julian did not work on the Sabbath, but when it ended he was recharged for the next week's combat. If Malthus is in heaven, he's in for an argument, laced with facts, facts, facts.

* * *

The present book, *Hoodwinking the Nation: Fact and Fiction about Environment, Resources, and Population,* is the first of Julian's work to be published posthumously. In it Julian goes after the media-academic complex that he believed purposefully, organically, structurally, and ideologically purveys "false bad news," specifically about the environment, resources, and population. As someone who some years ago wrote a book titled *The Good News Is the Bad News Is Wrong,* I certify that the good news is not only that Julian got it right, as you might expect, but that he did so with originality and flair.

Hoodwinking the Nation is a devastating critique, suffused with the outrage that so often served as Julian's trademark. The general reader will find it lively and informative, sometimes startling, and often fun—unless of course the general reader in question is Albert Gore. Suffice it to say that the vice president, the author of *Earth in the Balance,* whom Simon calls "the Hoodwinker-in-Chief," does not, uh, come off well in this volume.

Julian loved to make bets. Fifty years from now readers who peruse *Earth in the Balance* by Albert Gore and *Hoodwinking the Nation* by Julian Simon will giggle at one of them. Let's bet which.

BEN J. WATTENBERG
American Enterprise Institute

An Apologetic Preface

Its inconsistencies make this book a style editor's nightmare. The chapters vary greatly in length. The analytic techniques differ from chapter to chapter, spanning the range from a case study to compilations of statistical data. The prose varies from chatty to academic-ese cum footnotes.

The reason for this regrettable but unavoidable variety is that I am trying to describe and analyze an entire elephant of an issue, and the different parts of an elephant need very different treatments: feet require less space to discuss than do heart and brain; genes require electron microscopy whereas tusks require tape measures. A less-fanciful analogy is a rounded discussion of the causes of a plague: biology, medical preventive measures, therapies, economics, sociology, psychology, and politics are part of the analysis, and all require different sorts of discussion.

Similar treatment is appropriate for similar subjects; one can write 20 similarly formatted chapters on 20 sopranos or 20 skyscrapers or 20 insurrections. But similar treatment does not work for very different dimensions of a single multifaceted phenomenon such as false bad news about the environment, resources, and population growth.

Shifting to a different metaphor: Someone asked Abraham Lincoln, "How long should a man's legs be?" Lincoln answered, "Long enough to reach the ground." I hope that each chapter is long enough to reach the ground, and no longer.

Many of the examples of news stories and public events used here are not recent. I have been collecting material for this book, and writing these essays, for most of two decades, and the particular examples date from when I was thinking about that particular topic. This may seem a drawback, but it illustrates one of the themes of the book: up-to-dateness and true "news" are inconsistent with a long historical perspective, and therefore are often inconsistent with a valid set of facts and a sound understanding of events and trends.

Various chapters draw heavily upon my earlier writings; the book therefore may be thought of as part of an evolutionary process in knowledge development, rather than invention *de novo*. I make no apologies for this. Few of my readers are likely to have read much of the earlier work. Just as with a play or a piece of music, it would seem, there is nothing sacrilegious or indecent or exploitive about putting the same material before new audiences.

I hope that the book leads you to better understand the subject at hand despite all these aesthetic flaws.

Introduction

Public-opinion surveys tell us this for sure: most persons in the United States believe that our environment is getting dirtier, we are running out of natural resources, and population growth in the world is a burden and a threat. Ten minutes spent questioning grade-school children will confirm that even our youngest citizens hold these beliefs.

It also is sure by now that these beliefs are entirely wrong. Though it is not well-known to the public, there is broad scientific consensus that the air and water in the United States are getting cleaner rather than dirtier, that natural resources are becoming less scarce rather than more scarce, and that there is no quantitative evidence that population growth is detrimental to economic growth in poor countries or rich ones.

Why is there so much false bad news about the subjects of the environment, resources, and population? From the very first public talk about population growth that I gave in 1969, and the first article that I wrote for the broad public in *Science* in 1980, this question has arisen again and again from the few people who took the argument seriously.

An even tougher question is this one: Why do we *believe* so much false bad news about the environment, resources, and population? What we read and hear would not matter unless we also come to believe that "news" is true. Hence this book is about a complex structure:

- the *nature* of the false bad news;
- the *production* of false bad news by researchers, politicians, organizations;
- the *dissemination* of it by the press and television; and
- our propensities as human beings that lead us to *consume* (and be consumed by) that body of false statements.

In the 1980 *Science* article I gave this answer to the question, "Why do false statements of bad news dominate public discussion of these topics?"

- There is a funding incentive for scholars and institutions to produce bad news about population, resources, and the environment. The AID and the UN's Fund for Population Activities disburse more than $100 million each year to bring about fertility decline. Much of this money goes to studies and publications that show why fertility decline is a good thing. There are no organizations that fund studies having the opposite aim.
- Bad news sells books, newspapers, and magazines; good news is not half so interesting. Is it a wonder that there are lots of bad-news bestsellers warning about pollution, population growth, and natural-resource depletion but none telling us the facts about improvement?
- There are a host of possible psychological explanations for this phenomenon about which I am reluctant to speculate. But these two seem reasonably sure: (1) Many people have a propensity to compare the present and the future with an ideal state of affairs rather than with the past or with some other feasible state; the present and future inevitably look bad in such a comparison. (2) The cumulative nature of exponential growth models has the power to seduce and bewitch.
- Some people publicize dire predictions in the idealistic belief that such warnings can mobilize institutions and individuals to make things even better; they think that nothing bad can come of such prophecies. But we should not shrug off false bad news as harmless exaggeration. There will be a loss of credibility for real threats as they arise and a loss of public trust in public communication. As Philip Handler, president of the National Academy of Sciences, testified to congressmen in the midst of the environmental panic of 1970: "The nations of the world may yet pay a dreadful price for the public behavior of scientists who depart from . . . fact to indulge . . . in hyperbole" (Simon 1980).

Since then I have added to the list of possible explanations:

- Many people prefer bucolic surroundings to resource development.
- We may carry psychological propensities deep in our psyche that predispose us to warnings of doom.
- Journalists and interest groups use marvelously evocative inflammatory rhetoric to arouse fear—"population bomb," "empty pumps," "save the children," "end of the world as we know it," and "end of the age of affluence."
- Simple racism may also play a role, especially with respect to population growth in other parts of the world, and with respect to immigrants of various shades and ethnicities entering the United States.
- Some activists display an attitude toward the facts that induces them to exaggerate and even lie when they are convinced that the eleventh-hour danger to the public justifies such dishonest practices. And joining the environmental movement is seen by many as a last chance to do good, just as joining the Communist Party in the 1930s seemed an opportunity for social contribution by many generous-minded people.
- A set of unsound *ideas* undergird the newspaper and television stories and provide the intellectual infrastructure that give these stories credibility. These ideas fall into two categories: misunderstandings of the nature of resource creation and population economics, and misunderstandings of the nature of a modern complex social-economic system.

This book is the culmination and the integration of those individual studies.

What Does This Book Contain?

Following this introduction, the first chapter describes the facts about population growth, natural resources, and the environment and then presents survey evidence on the nature of beliefs held by the public on the same topics. The discrepancy between the facts and public beliefs sets up the puzzle that the remaining chapters in the book attempt to explain.

Chapters 2 through 5 explain how and why false bad news is produced. Chapter 2 describes the role of government in producing environmental news scams. It presents a case study of the scare created by a government report about how our farmland is vanishing because of the increased rate of urbanization. Chapter 3 discusses the intellectual bases of concepts that lead to scares about resource depletion and population growth. Chapter 4 describes organizational features of intellectual life that result in the dominance of doomsday analyses. Much of the alarm about Environment, Resource, and Population topics originates with biologists, as has been the case for centuries. Chapter 5 discusses what is special about the thinking of biologists that makes many of them become so alarmed about these topics.

Chapters 6 through 8 describe the processes by which false bad news is disseminated and their importance. Chapter 6 describes how journalists do not receive appropriate training in statistics and science and thus gather data in ways that lead to inaccurate conclusions. Chapter 7 illustrates the misuse of statistics in books by Bill Bennett and Al Gore and articles in the *Washington Post*. Chapter 8 suggests the consequence of poor journalistic practices is a mal-informed public.

Chapters 9 and 10 discuss the psychological and cultural mechanisms that make people receptive to bad rather than good news. Chapter 9 discusses reference points used to make comparisons: why we seem to think that the glass is half empty rather than half full and why most people have views of the past that are too positive and views of the present that are too negative. Chapter 10 expands on positive views of the past by discussing the role of future doom in religious and environmental thought.

I do not promise you a complete and coherent explanation of why we hear so much false bad news. The book inevitably will leave you with open questions about the relative importance of the various factors discussed, and how they interact; it does not have the systematic, satisfying wholeness of a manual explaining the workings of the internal combustion engine. There may even be important elements in operation that I do not discuss. I can promise you, however, that reading the book will enable you to appreciate that the overabundance of false bad news is a very complex phenomenon that has roots in our psyches, our economic system and the incentives it provides, the history of the past few decades, and a host of other factors that influence why some organizations supply, and many individuals demand, the scary misinformation.

Also, please do not expect a shivery-delicious orgy of press bashing; though there is a fair amount about the press and other institutions (and a bit of blood and gore), the core of the book is its analysis of the intellectual and emotional processes of thinking, rhetoric, and belief.

The intention of presenting the ugly stories is not to inveigh against immorality. Self-righteous indictments of miscreants solve nothing because the problem is a structural one: there are too few institutional sanctions against the undesirable practices; there are too few instruments that punish the unethical behavior. To get more people acting with greater integrity we need to strengthen the structure; as some wise person with balanced judgment observed long ago, locks help keep honest persons honest. The problem is that there are few obvious ways to improve the structure in a free society. We should also note, however, the many individuals who refrain from grabbing the short-run gains from peddling false bad news and instead work and live with integrity.

Some alert journalists have written about something that they always knew—the propensity of commercial media to trumpet bad news and put a bad-news spin on even good news, and they have discussed this propensity in the press. Other journalists have become aware that our material world—including the environment—is in better shape than ever before in human history, and have written articles to that effect. Those journalistic accounts, however, are mostly informed opinions—correct informed opinions, and therefore valuable and to be appreciated for that reason—but still opinions. In contrast, this book purports to be scientific scholarship about those same issues. That is, it adduces empirical data, provides systematic examination of prior studies, and searches the scientific literature for scholarly analyses of the issues, though conveying the findings in as lively a fashion as possible. I therefore hope that the book establishes a more solid basis for future discussion of these issues.

1. What Do Americans Wrongly Believe about Environment, Resources, and Population?

The Factual Context

Until the 18th century there was slow growth in population, almost no increase in health or decrease in mortality, slow growth in the availability of natural resources (but not increased scarcity), increase in wealth for a few, and mixed impacts on the environment. Since then there has been rapid growth in population because of spectacular decreases in the death rate, rapid growth in resources, widespread increases in wealth, and an unprecedentedly clean and beautiful living environment in the richer capitalistic countries along with a degraded environment in the poor and socialist countries.

The increase in the world's population represents our victory over death. In the 19th century the earth could sustain only one billion people. Ten thousand years ago, only one million could keep themselves alive. Now, five billion people are on average living longer and more healthily than ever before.

The current gloom-and-doom about an environmental crisis is all wrong on the scientific facts. Even the Environmental Protection Agency acknowledges that U.S. air and water have been getting cleaner rather than dirtier in the past few decades. Every agricultural economist knows that the world's population has been eating ever better since World War II, defying simplistic Malthusian reasoning. Every resource economist knows that all natural resources have been getting more available rather than more scarce, as shown by their falling prices over the decades and centuries. And every demographer knows that the death rate has been falling all over the world;

life expectancy almost tripled in the rich countries in the past two centuries and almost doubled in the poor countries in just the past four decades. This is the most important and amazing demographic fact—the greatest human achievement in history. It took thousands of years to increase life expectancy at birth from just over 20 years to the high 20s about 1750. Suddenly, about 1750, life expectancy in the richest countries began to rise so that the length of life that could be expected for a baby or an adult in the advanced countries jumped from less than 30 years to perhaps 75 years. Then starting well after World War II, the length of life that could be expected in the poor countries leaped upwards by perhaps 15 or even 20 years because of advances in agriculture, sanitation, and medicine. It is this decrease in the death rate that has caused there to be a larger world population nowadays than in former times.

Moreover, the evidence clearly indicates that population growth does not hinder economic development, and all the statistical studies show that faster population growth does not cause slower economic growth. In the 1980s there was a complete reversal in the consensus of thinking of population economists about the effects of increased population. In 1986, the National Research Council of the National Academy of Sciences completely overturned the "official" view away from the earlier worried view expressed in 1971. It noted the absence of any statistical evidence of a negative connection between population increase and economic growth and said, "The scarcity of exhaustible resources is at most a minor restraint on economic growth."

For proper understanding of the important aspects of an economy we should look at the long-run trends. Almost every long-run trend in material human welfare points in a positive direction, as long as we view the matter over a reasonably long period of time. And there is no persuasive reason to believe that these trends will not continue indefinitely. But the short-run comparisons—between the sexes, age groups, races, and political groups, which are usually purely relative—make more news.

Would I bet that the long-run trends have been the "real" trends and that they will continue into the future? Certainly. I'll bet a week's or month's pay that just about any trend pertaining to material human welfare will improve rather than get worse. You pick the comparison and the year. (Anything I win goes to pay for more research.)

Public's Beliefs Contrast with the Facts

Let's lay the factual groundwork for what the public wrongly believes about environment, resources, and population. These are some of the main subjects of false bad news and my own particular interest.

An authoritative wide-ranging review of international public opinion data in *Science* concludes that "people in both developing and industrial countries perceive that environmental quality has been and is continuing to worsen" (Bloom 1995, 357). That's the general state of the matter (though the word "perceive" wrongly suggests that what people think about the matter really exists, which is incorrect).

Late in the 1980s, disposable diapers became a cause célèbre. Government agencies have used the estimate that disposable diapers account for 12 percent of total trash. A poll of attendees at a National Audubon Society meeting produced an average estimate that diapers account for 25 percent to 45 percent of the volume of landfills. And a Roper poll found that 41 percent of Americans cited disposable diapers as a major cause of waste disposal problems."[1]

That's what the public (and a government agency) *think* is true. Yet according to the best available estimate, the diapers constitute "no more than one percent by weight of the average landfill's total solid-waste contents . . . and an average of no more than 1.4 percent of the contents by volume."[2]

The public also misunderstands the extent of the waste from fast-food packaging. The Audubon Society meeting poll found

[1]Rathje and Murphy, 1992, p. 162.

[2]Ibid.

an average estimate of 20 to 30 percent of landfills; the actual volume is "no more than one-third of 1 percent."[3]

In school, 46 percent of children aged 6–17 said they had heard about the importance of "solid waste" disposal (36 percent recycling, 15 percent litter, 6 percent garbage/landfills) in school in the 1991–92 school year.[4] (One wonders what the comparable numbers would be for the importance of honesty, hard work, and the free enterprise system.)

The propaganda to the children is effective. A "national survey of children ages 5 through 8 asked these questions: 'What would you do to make your city a better place?' and '. . . America a better place?' A majority of the kids . . . answered 'Clean up.' " This finding is significant for what the children *do not* say. There is no mention of "Build schools and parks," or "Go to the moon," or "Help those who are less well-off."

Nor is this childish thinking confined to children. An attorney for the Environmental Protection Agency complained to the newspaper that the post office's new self-sticking stamps were "long-term environmental mistakes," because "as everyone knows, there is a solid waste problem in the country," and stamps with plastic "strike me as an incredibly irresponsible use of our limited petroleum resources."[5] The more relevant limited resource is newspaper space, and the editor's choosing to print this letter rather than the hundreds of other contenders on serious subjects strikes *me* as "irresponsible."

And consider energy as a "problem." The percentage of the public that said energy is the "most important problem facing the nation" jumped from 3 percent in September 1973, to 34 percent in January 1974, and then quickly fell back down to 4 percent. Then the rate went sharply up and then down again in 1977, and then up once more in the summer of 1979,

[3]Rathje and Murphy, pp. 114–5.

[4]Environmental Research Associates, *The Environmental Report: The Power of Children*, reported in *Public Opinion*, U.S. Council for Energy Awareness, November 1992.

[5]*The Washington Post*, November 25, 1989, p. A22.

following the price rise by OPEC, and then down once more.[6] Fully 82 percent in 1979 said that "the energy situation in the United States" is "very serious" or "fairly serious,"[7] but another series of polls also showed that public concern quickly dropped in the 1980s. These swift changes in the public's thinking illustrate the volatility of concern about energy and oil.

And it's not just diapers, packaging, and energy. People are frightened about the entire range of environment-resources-population issues. According to a CBS News Survey before Earth Day 1990, "The American public has an almost doomsday feeling about the national seriousness of environmental problems."[8] Across the board, the public, the environmental organizations, and the press say that pollution in the United States and the world is not just bad, it is getting worse. Public statements on the matter emanate from prominent scientists, politicians of every stripe, and religious leaders of every denomination. For Earth Day in 1994, the National Council of Churches and something called the National Religious Partnership for the Environment distributed tens of thousands of "Environmental Awareness Kits" to churches and synagogues. For Protestant churches the following was part of the recommended program (Sirico 1995):

> Minister: We use more than our share of the Earth's resources. We are responsible for massive pollution of earth, water and sky. . . . We thoughtlessly drop garbage around our homes, schools, churches, places of work, and places of play. . . . We squander resources on technologies of destruction. Bombs come before bread.
>
> Congregation: We are killing the earth. . . . We are killing the waters. . . . We are killing the skies.

Notice that there was nary a word in that declaration about the creating and building that we do—and which obviously

[6] *Advertising Age*, September 24, 1979, p. 48.

[7] *Champaign-Urbana News Gazette*, October 16, 1977, p. 2-A.

[8] April 16, 1990.

is larger than the destruction, because our world becomes healthier, safer, and wealthier with every passing decade.

You want amity and agreement among the religions? You've got it here. Protestants (including evangelicals), Catholics, and Jews (Reform and Conservative) joined in the NCC campaign. In 1991, even the nation's Roman Catholic bishops "acknowledged that overpopulation drains world resources." They asked Catholics "to examine our lifestyles, behaviors and policies, to see how we contribute to the destruction or neglect of the environment." The pope issued a 1987 encyclical, "Sollicitudo Rei Socialis," and a 1990 New Year's message on this theme of environmental "crisis" and "plundering of natural resources," and "the reality of an innumerable multitude of people." (The pope apparently has "gotten religion" since then and turned around on the issue.)

The environmentalist ideal has suffused the Jewish community, too. In Washington[9] there was held a "Consultation on the Environment and Jewish Life," intended as "a Jewish communal response to the world environmental crisis." The signers of the invitation included just about every big gun in the organized Jewish community. The invitation letter said: "We appreciate the many important issues on the Jewish communal agenda. But the threat of ecological catastrophe is so frightening and universal that we believe we must mobilize our community's considerable intellectual and organizational resources as soon as possible."

Just about all of these assertions of rising pollution are nonsense. But they are dangerous nonsense. Not only is the public frightened, but public concern is *increasing*, as several types of public opinion polls confirm.

Are Things Getting Worse or Better?

Many more people believe that trends in various indicators are deteriorating.

[9]March 9 and 10, 1992.

- "Would you say that people are subject to more risk today than they were 20 years ago, less risk today, or about the same amount of risk today as 20 years ago?" Seventy-six percent of the public said "more risk" and only 6 percent said "less risk."[10]
- A 1988 survey found that "eight in ten Americans (81%) were convinced that 'the environment today is less healthful than the environment in which my parents lived.' "[11]
- In 1990, 64 percent said that pollution had increased in the past 10 years, while 13 percent said it had decreased.[12]
- "Compared to twenty years ago, do you think the air you breathe is cleaner today, or more polluted?" Six percent said "cleaner," and 75 percent said "more polluted." With respect to the "water in the lakes, rivers, and streams," 8 percent said "cleaner" and 80 percent said "more polluted."[13]
- In 1991, 66 percent of Americans responded "worse" to "Overall, do you feel the environment has gotten better, gotten worse, or stayed the same over the past 20 years?" Only 20 percent said "better."[14]
- In 1996, people in New Zealand—a country that then had a flourishing economy, a peaceful society, and extraordinary beauty—were asked to agree or disagree with the statement "It is getting harder and harder for the ordinary family to give their children a good future." The results were: "strongly agree," 63 percent; "agree," 84 percent; "strongly disagree," 3 percent; "disagree," 10 percent (*TV3/CM Research Gallup Poll*).
- When asked "Looking ahead to 20 years from now, do you think the risks to society stemming from various

[10]Harris 1980, 9.

[11]Shapiro, 1991.

[12]MG/AP poll 31, May 11–20, 1990.

[13]These polls were taken in the midst of the Earth Day publicity, however. *New York Times*/CBS, March 30–April 2, 1990.

[14]*Wall Street Journal*, August 2, 1991, A1. For more data, see Dunlap and Scarce (1991).

scientific and technological advancements will be somewhat greater, somewhat less, or about the same as they are today?" 55 percent of the public said "somewhat greater," and 18 percent said "somewhat less" (Harris 1980, 11).

Trends in Perceptions about Pollution

The proportions of people expressing worry about pollution problems show large increases over recent years.

- In Harris polls the proportion who said that air pollution by vehicles was "very serious" rose from 33 percent in 1982 to 59 percent in 1990.[15]
- The proportion who said that "air pollution from acid rain, caused by sulfur dioxide emissions from power plants" was "very serious" rose from 42 percent in 1986 to 64 percent in 1990.[16]
- There was an increase from 30 percent in 1986 to 49 percent in 1990 saying "very serious" for "air pollution by coal-burning electric power plants."[17]
- However, a 1991 Roper poll found that people thought that the environment would be cleaner five years later than at the poll date, unlike a similar comparison in 1980. And people's assessment of the environment "at the present time" was less positive in 1991 than in 1980.[18]
- In 1990, 44 percent said they "expect pollution to increase," and 33 percent expected it to decrease.[19]
- A survey of high school students found that "the only interviewees who didn't share the perspective . . . that the environment is going to be destroyed completely . . . were the worst educated of the inner-city youth."[20]

[15] *The Harris Poll,* April 1, 1990.

[16] *The Harris Poll,* April 1, 1990.

[17] *The Harris Poll,* April 1, 1990.

[18] *Roper Reports* 92–1, p. 25.

[19] MG/AP poll 31, May 11–20, 1990.

[20] *Fortune,* March 26, 1990, p. 226.

Education and Environmental Awareness

One might wonder whether less-well-educated persons are less responsive to environmental issues simply because they know less. But surveys that ask whether "pollution increased in the past 10 years," or "decreased," or "stayed about the same" show that answers are not related to amount of education.[21] Education in large quantities would seem to increase one's propensity to rely on such abstractions.

A Sunday newspaper kids' page article purveys such bits of "obvious wisdom" as "It takes more than 500,000 trees to make the newspapers that Americans read on Sunday ... we're running out of places to put it . . . there aren't very many new places to put [landfills]."[22] The children are not told that trees are grown, and forests are created, in order to make newspaper.

Even grammar school texts and children's books fill young minds with unsupported assertions that humankind is a destroyer rather than a creator of the environment. It is not surprising that the consensus view of an informal *Fortune* survey of high schoolers on this "issue on which almost everyone agreed" was: "If we continue at the pace we're going at now, the environment is going to be destroyed completely."[23] A 1992 poll found that 47 percent of a sample of 6-to-17-year-olds said that "environment" is among the "biggest problems in our country these days"; 12 percent mentioned "economy" as a far-distant runner-up. Compare the almost opposite results for their parents: 13 percent responded "environment" versus 56 percent "economy."[24]

[21]The only striking difference is that women were more likely than men to say "increased" and less likely to say "decreased"—72 percent versus 56 percent and 8 percent versus 19 percent, respectively. There also was a slight gradient downward in "increased" with older groups (MG/AP poll 31, May 11–20, 1990).

[22]O'Neill, 1991.

[23]Sherman, 1990, p. 226.

[24]Environmental Research Associates, *The Environmental Report: The Power of Children,* reported in *Public Opinion,* U.S. Council for Energy Awareness, November 1992.

A surprise to me is that the aged are even more convinced than the general public—86 percent to 78 percent—that "people are subject to more risk today than they were 20 years ago." I would have thought that over their long lives, people aged 65 and over would have learned how much safer life is, as measured by rates of mortality and injury. Only 2 percent answered "less" to the above question, whereas the rate in the public as a whole was 6 percent (Harris 1980, 9).

2. The Vanishing Farmland Scam

The vanishing farmland scam is a crystal-clear example of concerted false scaremongering in which the perverse roles of the federal government, environmental organizations, and the press and television are undeniable and inarguable. Even the original purveyors of the false facts now agree that the widely reported scare was without foundation.

In 1980 typical headlines announced the existence of a farmland crisis: "The Peril of Vanishing Farmlands" (The *New York Times*); "Farmland Losses Could End U.S. Food Exports" (*Chicago Tribune*); "Vanishing Farmlands: Selling Out the Soil" (*Saturday Review*); and "As World Needs Food, U.S. Keeps Losing Soil to Land Developers" (*Wall Street Journal*).

The "crisis"? The urbanization-of-farmland rate supposedly had jumped by a multiple of 3 from the 1960s to the 1970s, from less than one million acres per year to three million acres per year. The entire "crisis" was hokum, however.

This was not a regrettable but understandable exaggeration of a real problem, but a nonproblem manufactured out of whole cloth by the Department of Agriculture and some members of Congress under the guise of concern about food production for the starving world. The crisis was created for the benefit of environmentalists and people who own homes that abut areas that might be developed into housing developments, and whose vistas and ambience might thereby be affected.[1]

The Origins of the Scam

When newspaper and television stories about "vanishing farmlands" first began to appear around 1979, I was living in

[1]The connection between the farmland scare and prevention of housing construction has been documented for California by Frieden (1979).

Champaign-Urbana, Illinois, the heart of the Illinois farm belt. The front page of the local paper, the *News-Gazette*, carried the same scare stories about the world's food supply that deliciously agitated the intellectual trendsetters in Washington and New York. The farm page, however, printed exactly the opposite news—agricultural prices were falling and production was increasing all over the world. The Champaign-Urbana paper, and other papers in agricultural states, got the facts straight (at least on the farm page!) because those facts were directly relevant to the farmers in the area, who could see the trends when they went to market. Farmers were worrying about too much food production rather than too little when they made their planting decisions each year.

The issue caught my attention because my research has shown that population growth is a benefit rather than a barrier to world development in the long run. The apparently impending shortage of farmland was thrown at me as a counterargument, along with the related assertions that the world was in danger of running out of copper, oil, water, and other natural resources.

Even without any specific research regarding its assertions, the vanishing-farmland assertions of the USDA and the National Agricultural Lands Study (NALS) seemed likely to be wrong. First, large changes, like the tripling of the rate of farmland conversion, usually do not occur rapidly in major sectors of our economy. The report of any sharp change is immediately suspect. Second, each and every one of the previous population-growth doomsday scares had turned out to be spurious. Third, local Champaign-Urbana farmland-preservation enthusiasts—the environmental movement was particularly strong in the area because of the presence of university faculty and students, especially the biologists and political scientists—were campaigning to prevent pieces of the county's farmland from being converted to a shopping mall, an industrial complex, and even second farmhouses for farm families' grown children. The numbers that these persons were publicizing would have been laughable if they were not being used toward a serious purpose and in a politically effective manner. My research began with those numbers.

The Champaign County Soil and Water Conservation District, the federal Farmers Home Administration, and the editorial page of the Champagne Urbana *News-Gazette* asserted that 30,000 acres of county cropland had been urbanized from 1960 to 1978, an average of 1,667 acres per year. But as of 1978, the entire urban area of the county—the city of Champaign-Urbana plus the only village, Savoy—totaled only 18,695 acres, according to aerial maps. Those maps also showed an average increase in urbanized areas of only about 320 acres a year, a far cry from the publicized 1,667 acres. Soil Conservation Service data indicated that an average of 538 acres of farmland per year was urbanized from 1958 to 1967, and there was no obvious reason why the rate would have increased. And most important, any casual motorist in the county could see the impossibility of almost 10 percent of the county's 640,000 acres *ever* having been paved, let alone that much since 1960. Furthermore, standard data from the Census of Agriculture showed that most of the county was being farmed.

A few phone calls established that the 30,000-acres number originated with the newsletter of the Illinois Department of Agriculture, and that same newsletter carried other numbers that were obviously preposterous. One farming county supposedly had lost 47 percent of its farmland between 1960 and 1978. After a bit of prodding, department of agriculture officials acknowledged that the newsletter numbers were "grossly inaccurate." But the people publicizing the numbers had never bothered to check their data even though the data defied belief. One can only assume that the numbers were used so uncritically—by the editorial writers of the *News-Gazette* and others—simply because they fitted preconceptions and values.

The Scam Unravels

To make a very long story very short, several scholars—including William Fischel of Dartmouth; Clifford Luttrell of the St. Louis Federal Reserve Bank; Emery Castle, then president of Resources for the Future; John Fraser Hart of the University of Minnesota; and I—began to dig into the data. We all found that the three-million-acres-a-year rate was most implausible in light of various sets of data from other sources and given the nature of the surveys from which the NALS estimate was drawn.

Shortly, we began to get a great deal of help from H. Thomas Frey, a geographer who had been the keeper of the urbanization and other land-use data for the Economic Research Service of the USDA for many years. Tom could find no support for the scary new numbers in the standard set of sources from which he yearly distilled his estimates, and he had no reason to believe that the rate had increased from the rather constant rates over several previous decades. So he kept us alerted to useful pieces of information as they appeared.

Everyone agreed that in 1967 the total urban and built-up area in the United States (excluding highways, railroads, and airports) was between 31 and 35 million acres. It was also agreed that the rate of urbanization was slower in the 1960s than in the 1950s. Yet NALS said that over the 10 years from 1967 to 1977, there was a 29-million-acre increase in urban and built-up land.

That is, over the course of more than two centuries, in the process of reaching a population of about 200 million people, the United States built towns on 31 to 35 million acres. NALS asserted that suddenly in the course of another 10 years, and with a population increase of only 18 million people, the acreage of urban and built-up area almost doubled.

To put it differently, the long-run trend in the decades up to 1970 was an annual increase of about one million acres of total land urbanized per year. In addition, this rate of increase was constant or slowing down. In contrast, the Soil Conservation Service, in conjunction with NALS, asserted that the rate jumped to two to three million acres yearly from 1967 to 1975 or 1977 (depending on which version you read).

It is hard to understand how thoughtful journalists, let alone USDA researchers, could believe the NALS scenario. But believe it they did—or at least they promulgated the scenario widely. This credulity belies the picture of journalists as hard-boiled skeptics.

Tom Frey also began to have a personal problem. He began this episode as an amiable organization man who went through channels, did everything by the book, and was accommodating whenever he could be. But the widely publicized estimate did not square with the facts as he knew them, and he so informed his superiors. They systematically bypassed and ignored his assessment at each level all the way to the secretary of agriculture, Bob Bergland.

As time passed, Frey became more outspoken even as the heat upon him intensified and even though no one else in the USDA would speak up for his position. The pressure was always indirect, but its presence was sufficiently unmistakable that Frey worried about his future. He was forced into confrontation with Michael Brewer, the research director of NALS. Consequently, his official reports were altered to the extent that he refused to put his name on them.

Still, the more pressure his superiors put on him, the more strongly Frey insisted that his estimates were correct, and he distributed his estimates to those he felt would make use of them. He hardened into an authentic hero. He was not a whistle-blower in the conventional sense because he never "went public" himself. But he did his job effectively and with courage, the sort of public servant that citizens deserve but whose existence we sometimes doubt.

There were two bases given for the publicized three-million-acre number; NALS shifted from one to the other when either was criticized: (1) a small-sample 1975 resurvey of part of the 1967 sample "inventory" of farms, done by the Soil Conservation Service (a similar inventory had been done in 1958), and (2) the 1977 sample inventory. Seymour Sudman, an expert in research design, joined me in a technical analysis showing that there were so many flaws in both the 1975 resurvey and the 1977 survey that both should have been considered totally unreliable. The flaws included an incredible error that put the right numbers in the wrong columns for large areas of Florida.

Though I talk about "bases" for the NALS estimate, I may be giving them too much dignity. The three-million-acre figure was presented in a booklet entitled "Where Have the Farmlands Gone?"—of which 500,000 copies were sent out—weeks before the NALS "study" was even published in the fall of 1979. Everything else that NALS did may be seen as an effort to defend its initial position. This would seem to have been a choice bit of business for the press to have exposed—but it never happened.

Politicized Responses

Now under fire, NALS mobilized help to fudge its gross miscalculations. The coverup started with the governor of Illinois's issuing an executive order citing "a loss of 100,000 agricultural acres every year." Dale McClaren of the Greater Wabash Regional Planning Commission brought this to my attention, and together we wrote in the *Wall Street Journal* showing that on the basis of Census Bureau data there was no decline at all, but rather an increase. The Illinois administration, in conjunction with NALS, got the Census Bureau to issue "corrected" data for the state.

USDA had claimed that farmland was decreasing. But we showed that farmland was in fact increasing. The *Journal of Soil and Water Conservation* then wrote as follows about the aftermath:

> The new information [that Illinois farmland was increasing rather than decreasing] came out of a feud between the Illinois Department of Agriculture [then headed by John Block who became Secretary of Agriculture under Ronald Reagan] and Julian Simon . . . (Cook 1981).

My assertions did not sit so well with Illinois agriculture officials. They decided to contact the Bureau of the Census in Washington because they could not believe that land in farms had increased. The bureau responded that the 1969 and 1974 censuses had underenumerated land in farms.

That is, the Census Bureau now said that its earlier acreage estimates for Illinois should have been higher and thus there could have been an actual decline even though the record showed an increase. The "revised figures showed that between 1974 and 1978 Illinois' land in farms had declined by 425,000 acres. The earlier figures had shown a 639,000 acre increase" (Cook 1981).

NALS then got the Census Bureau to produce a similar adjustment for the United States as a whole. "The result: The latest data show a national decline of 88 million acres in land in farms between 1969 and 1978—an annual rate of 9.8 million acres," wrote *The Journal of Soil and Water Conservation* (Cook 1981).

Analysis of the adjustment showed it to be as full of holes as Swiss cheese. And eventually the Census of Agriculture revealed detailed data on the appropriate adjustment showing that land in Illinois farms and in cropland had, as we said, indeed increased from 1974 to 1978.

Ironically, John Block ran into trouble because of his belief that farmland was becoming more scarce. "Financing arrangements used by Agriculture Secretary John R. Block to prop up his multimillion-dollar farming empire, apparently shaken by falling land values . . . are raising questions on Capitol Hill." His banker explained: "We thought we were going to have to feed the world."[2]

Later Block became president of the National Wholesale Grocers' Association, which has a stake in cheaper food production. And on July 16, 1988, Block wrote in the *Washington Post* criticizing Lester Brown, who was and is one of the great proponents of the NALS point of view: "In 1980, respected agricultural experts were predicting an impending food shortage and possible mass starvation . . . but the suggestion that the world [now] stands on the ragged edge of food shortages is preposterous. . . . I don't buy the suggestion that the world is running out of productive land." Better late than never.

Somewhere along the line the NALS principals had a falling out. In the press, NALS research director Michael Brewer accused NALS head man Robert Gray of inflating the key

[2]*Washington Post*, June 3, 1984, p. A3.

estimate for political purposes. By the time NALS closed down, Gray and Brewer were no longer speaking to each other because of this and related disagreements.

After several articles appeared in technical journals as well as in policy journals such as *The Public Interest* and *The American Spectator*, the urbanization-of-farmland scare seemed to die down a bit, but not before the private American Farmland Trust was organized in 1980 by former employees of NALS. American Farmland Trust spends a couple of million dollars annually to "protect" the United States from the danger of vanishing farmland.

The Truth Officially Acknowledged

Now fast forward to 1984. The Soil Conservation Service issued a paper by Linda Lee of Oklahoma State University that completely reversed the earlier scare figures and confirmed the estimates by "our side." The accompanying USDA press release (April 10, 1984, kindly sent by Tom Frey) made it clear that the former estimates were now being retracted. "The acreage classified as urban and built-up land was 46.6 million acres in 1982, compared to 64.7 million acres reported in 1977." Please read that again. It means that whereas in 1977 the SCS had declared that 64.7 million acres had been "lost" to built-upon land, just five years later SCS admitted that the actual total was 46.6 million acres. That is, the 1977 estimate was fully *50 percent too high,* a truly amazing error for something so easy to check as the urbanized acreage of the United States.

With unusual candor, the USDA press release added,

> The 1982 data, which correlate closely with data from the 1980 U. S. Census of Population, [the census was not available at the time of the argument described above, but later fully corroborated Frey's estimates based on prior data] are considered accurate because of the availability of better maps, more time for data collection, many more sample points, and better quality control. The 1977 estimate thus appears to have been markedly overstated.

You might say so.

It is good that the USDA chose to state the situation forthrightly even though it meant acknowledging a mind-boggling amount of egg on its face. But how could the USDA earlier on have been that ingenuous—or that disingenuous? And how could it hold onto those estimates for several years in the face of persuasive criticism from several quarters?

The same USDA press release of April 10, 1984, contained a second quiet bombshell: Erosion has been *lessening rather than getting worse*, despite NALS claims to the contrary. Indeed, there is no evidence that loss of soil is a public problem, though of course it is a management problem for individual farmers just like maintenance of their farm equipment. But in this case, too, nothing appeared in print (to my knowledge) to make the public aware of this new nondanger and of how the public was misled.

The Response of the Press

Unfortunately the press did nothing to uncover the scam. Even the press release reversal and "confession" did not evoke coverage even though the original scare story was a front-page headliner for the *Chicago Tribune* and a cover story for news magazines.

After waiting some months for the coverage that never appeared, I decided to contact some newspapers. My across-the-alley neighbor, who worked for the *Wall Street Journal*, liked the idea, but the writer to whom he passed the material decided after several months to do nothing. I then got in touch with a reporter from the *New York Times* who works on related topics, but he was not interested. Nor the *Washington Post*. Nor the *National Journal*. Nor some others that I have forgotten. In short, nothing.

In late 1985 I got in touch with Gregg Easterbrook of the *Atlantic Monthly*. Editor Bill Whitworth had excerpted three chapters from a 1981 book of mine, and therefore I figured the *Atlantic* might be interested and sympathetic. Both found the story appealing, and Easterbrook's piece, "Vanishing Land Reappears," was in the July 1986 edition. It confirmed many of the facts mentioned here and described the political infighting involving NALS.

Reappearance of the Farmland "Crisis"

Did the farmland crisis then vanish for lack of factual support? Fat chance. The crisis seemed to have nine lives. A July 1986 newsletter from a group called Population-Environment Balance said that "the National Agricultural Lands Study projects that at the current rate of conversion, Florida will lose all its prime agricultural land by the year 2000." A story in the *New York Times* (July 14, 1987) began with "City sprawl, highways and other non-agricultural uses are taking American farmland at an annual rate that could involve acreage equal to the entire State of Missouri by the year 2030." The Fall 1987 issue of the Newsletter of Californians for Population Stabilization said, "The President and the Republican administration admit to the very conservative estimate that 2.1 million acres of agricultural land are paved over in the United States on a yearly basis. Experts in this field believe the true figure to be closer to 3 million acres per year." And the 1987 Annual Report of the American Farmland Trust said, "Between 1967 and 1982 alone, urban sprawl accounted for the loss of forty million acres of American farmland. Farmland conversion shows no signs of slowing."

The *Washington Post* ran some stories that described concerns in Maryland about farmland preservation:

> Increasing Development Of Md. Farmland Sparks Concern. Md. Farmland Disappearing At Rapid Pace. . . . The continued loss of farmland could force Marylanders to buy higher-priced goods shipped in from elsewhere (Nov. 6, 1988).
>
> "We have got this goal to save farmland so we can feed ourselves . . . ," said John Musselman, who heads the effort to preserve 20,000 acres of farmland in rapidly developing Howard County [Maryland] (Dec. 31, 1988).

Would anyone care to bet that this week some U.S. newspaper will not write again that three million acres are being paved over?

Did It Matter?

Did the false bad news matter? In 1980 Congress provided a tax break to owners who attach a "conservation easement" to their land that restricts development in perpetuity. Some states have programs to compensate the owners of conservation easements the difference between the market value before and after the easement. In 1981 the Farmland Protection Policy Act was enacted by Congress. Hundreds of state and local laws restricting farmland conversion also were passed. And the American Farmland Trust's 1985 Annual Report bragged that in that year "Congress adopted . . . a Conservation Reserve . . . conceived and championed by AFT" as part of the 1985 Farm Bill. All this legislation was based on wrong information, with no reasonable prospect of doing good for the nation, but with much prospect of causing harm to individuals and damage to the nation. So it goes in America.

A Postmortem

Free-lance journalist Julian Weiss (1983) gathered data for the Media Institute on how the print and electronic media handled the farmland-urbanization story. These were some of his findings:

- " 'We cultivated a good relationship with papers around the country,' declares Mr. Gray [NALS director]." Indeed they did. The first move was to mail their booklet with the three-million-acre figure to 1,800 newspapers—three copies to each newspaper. "Some material in the Study's 'package' was 'used verbatim' he says."
- Of the 80 journalists Weiss interviewed from February to May 1983, 72 (90 percent) still remembered NALS.
- "Sixty-three (86 percent of those who remembered it) felt that NALS influenced their own coverage of farmland conversion."
- "Even after dissidents from the Lands Study [members of the staff who told Gray that the publicized estimates were wrong] offered convincing evidence of the distortions contained in the final report . . . the media . . . with few exceptions [did not report] these charges."

Conclusion

Environmental reporters seem to believe reports about negative environmental trends issued by government agencies and environmentalist organizations and discount reports about positive trends in resources and environment. When shown the facts, these journalists usually say that even if cries of an environmental danger are somewhat overblown, they contain the germ of truth. I hope that the farmland case sobers them a bit. There was no grain of truth, only a bushelful of political deception and journalistic incompetence.

Unfortunately, fears about running out of metals, grains, water, and energy cannot be so easily shown to rest on faulty or "cooked" data, because there have been no public recantations by the sources of the erroneous information as there were with respect to farmland urbanization and soil erosion. But subsequent events have completely falsified the well-publicized 1970s gloomy prophecies of Paul Ehrlich, Lester Brown, Garrett Hardin, and Daniel Yergin. Knowledge-based increase in food productivity provides us ever more food per person on less and less cropland. The price of unprocessed food continues to fall, as it has throughout human history. Farmland prices have been falling rather than rising since the 1970s. Meanwhile, both the quantity of trees and the area devoted to recreation have been increasing, and people abroad eat better than ever. Moreover, this logic-defying process by which all good things increase at once can go on without limit, so far as we can tell—mind-boggling though that may be (Simon 1981/1996, 1982; Fischel 1985, chapter 1).

Yet the press continues to rely on the failed prophets and the discredited Malthusian ideas. The press and television have consistently purveyed a wrong-headed vision to the public of resource availability and the environmental condition. In so doing, they have aided and abetted government agencies and "environmentalist" organizations in scaring Americans about such nonproblems as the disappearance and erosion of farmland. Years after the original source retracted the alarming reports, the press and environmental groups continue to spread the false information.

I come not in anger, but in pain. Journalists take pride in their objectivity. But in reporting on population growth, natural resources, and the environment, objectivity goes out the window. The price in economic loss, misguided policies, and damage to national morale has yet to be calculated. But the costs may be fearfully high.

3. The Concepts That Lead to Scares about Resources and Population Growth

Why do political leaders tell us that life is more dangerous, our planet is "plundered" and "in crisis," we are running out of resources, and pollution is increasing, that is, that things are getting worse, when they are really getting better? Why do the politicians say that we need to "save the planet"? This chapter discusses some of the rhetorical devices used to communicate false messages of doom in prose, discourse, and politics.

The Seductiveness of the Malthusian Logic

The vision underlying the thinking of today's conventional writers about resources and population is the concept of fixity or finiteness of resources in the relevant system of discourse. This idea is found in Malthus, of course. But the idea probably has always been a staple of human thinking, because so much of our situation must sensibly be regarded as fixed in the short run—the number of bottles of beer in the refrigerator, the size of our paychecks, and the amount of energy that parents have to play basketball with their kids.

In contrast, the vision underlying sound thinking about resources—which is now the consensus vision of economists who study these subjects (National Research Council 1986)—is that it makes sense to treat the system operationally as *not fixed, rather than finite*. That is, a key difference between the thinking of those who worry about impending doom and those who see the prospects of a better life for more people in the future apparently is whether one thinks in closed-system or open-system terms. For example, those who worry that the second law of thermodynamics dooms us to eventual decline necessarily see our world as a closed system with respect to energy and entropy; those who view the relevant universe as unbounded view the second law of thermodynamics as irrelevant to this discussion. I am among those who view the relevant part of the physical and social universe as open for most purposes.

Which vision is better in the context of long-run decisions about resources and population is not subject to scientific test. Yet the choice profoundly affects our thinking.

Academics are particularly susceptible to the notion of Malthusian diminishing returns, perhaps because academics are more likely than laymen to believe in abstract theories. (Academics properly spend much of their lives battling to persuade others that abstract theorizing has importance and is not just an "ivory tower" recreation.) In my experience, journalists and businesspeople are less likely than academics to be taken with the simple Malthusian abstraction, perhaps because they have no professional stake in this idea (in contrast to many biologists and some economists) and perhaps because journalists are more attuned to reaching judgments and making decisions in light of the full richness of a situation—on their "intuition"—rather than on the basis of the logical relationships in a simple model. (More generally, businesspeople and newspeople seem to be more open to new ideas than academics, perhaps because a continuous flow of creative change is more crucial in their occupations.) Another element is the dead hand of expertise. As Kuznets tells us, "Experts are usually specialists skilled in, and hence bound to, traditional views; and they are, because of their knowledge of one field, likely to be cautious and unduly conservative" (Rosenberg 1972).

It is a puzzle why so many people—with biologists and physicists notable among them—are so sure that there *must* be some constraint to prevent humanity from growing both ever richer and ever more populous, and why theirs is the vision of unexpandable limits. One possible explanation is that all of us tend to bring our professional modes of thought to bear on other situations even if those modes are not appropriate to the situation at hand. For example, biologists liken the human population to an animal population and then apply the animal-ecology notion of "carrying capacity," though that notion is quite inapplicable to natural resources in a human context.

Another attraction of the closed-system vision is that the closure of the system enables one to use interesting mathematics, especially calculus and other optimization devices. From a purely physical point of view, a proposition about finiteness (or entropy) requires a bounded system. But where is the relevant boundary for our material world? Around the earth, excluding the sun? Around the earth plus sun plus solar system? Around other suns? Around a "universe" that may or may not be finite or expanding in the astronomer's eye? No boundary, no finiteness.

Still another root of the closed-system vision is the bewitching medieval notion of "first cause" or "ultimate cause," the idea that nothing happens which is not the result of other forces. And pushing back the causal sequence in an infinite regress, it seems as if there must have been an original causal force. This suggests a complete, and therefore closed, system.

For some, the closed-system vision arises because of a natural abhorrence of the loose-endedness of an open system. An interesting example of how this vision permeates our thinking: If you say that copper might be made of other metals, hearers say "alchemy." When you point out that nuclear bombardment transmutes metals, the hearers say "not practical," implying that it never *could* be practical. They may be correct. But there is no *logical* impossibility here. One can only be sure that something is impossible or impractical if one can be sure that the state of knowledge will not change in the future, that is, that capacities are limited because knowledge is limited. But isn't this just what people said in the past about the possibility of finding smaller constituent parts within the "fundamental" electron? And about the possibility of obtaining the vast amounts of energy that we get from a small pile of stuff called uranium?

There Can't Be a Free Lunch

"There is no free lunch," seems to imply that we have to pay for everything we get. This is another case of a good thought's going wrong by being applied to situations for which it was not designed. This slogan was originally intended to suggest that the *government* cannot supply free lunches to *all of us* and that no magic trick can increase our total national resources by passing laws and setting up bureaucracies; rather, we as taxpayers have to pay indirectly, sometime.

In other contexts, however, there *are* free (or below full-cost) lunches all the time. None of us always pays the full cost of production for what we get. In the modern world each generation gets its lunch at a lower cost of labor than did earlier generations, because earlier generations responded to their economic problems with ingenuity and energy. Our ancestors bequeathed us the intellectual wherewithal to get our lunch, if not entirely free, at least much cheaper than if we had to start from scratch. Compare what we "pay" to what Europeans had to "pay" for lunch and other meals a few hundred years ago. They paid most of every day's work, whereas we can buy the same amount of raw food with a small fraction of the work time it cost them. And there is no economic or physical force, and no concept in standard economic theory, that suggests that this progressive reduction in the cost of lunch cannot continue indefinitely. We eat our cheap lunch courtesy of the sweat of our ancestors' brows in mental and physical labor.

Lack of Historical Perspective

It is not surprising that most people are not aware that real prices of resources were higher in past years than now; this necessitates having knowledge of data back to (say) 1900 or 1800 and requires adjusting for inflation. Hence it is not surprising that views about impending resource scarcity are not informed by the contrary long-run trend of increasing availability.

It *should* surprise us, though, when mature, experienced journalists in high positions write that conditions are bad now without reference to how conditions were in the past. In 1980, columnist James Reston of the *New York Times* could write about "the civilized world that is now in such deep trouble," saying that "you can hardly pick up a paper these days without wondering what's wrong" and decrying our lack of leadership. Can this man have lived through the depression of the 1930s, Hitler, World War II, the Cold War, the Korean War, and the Vietnam War? And ex-senior editorial writer John Oakes of the same august newspaper reproduced the pessimistic findings of the *Global 2000 Report* almost word for word, like a press-conference handout. How could he have lived through the disastrous times of the past, when the environment was much more degraded and the materials more scarce, and yet write as if the world were headed straight toward doom?

The Long versus the Short Run

The distinction between the long run and the short run is crucial to the economics of population. In the developed world, additional people—babies or immigrants—are a burden in the short run. And focusing only on the short-run burden leads to a negative judgment about population growth. But in the long run, more people mean a higher standard of living for others. So the judgment about whether more people are good or bad economically depends on how one trades off the present versus the future. By most of my calculations, the discount rate would have to be quite high for additional people not to have a positive present value.

Furthermore, short-run costs are inevitable and obvious, whereas long-run benefits are hard to foresee. If your neighbor has another child, surely your school taxes will go up and there will be more noise in your neighborhood. And when the additional child first goes to work, per-worker income will be lower than otherwise, at least for awhile. It is, however, more difficult to foresee and to understand the possible long-run benefits. Because the increase in knowledge created by more people is nonmaterial, it is easy to overlook. Writers about population growth usually mention a greater number of mouths coming into the world, and sometimes note more pairs of hands, but never mention more brains arriving. This emphasis on physical consumption and production may be responsible for much unsound thinking and fear about population growth.

More People Create More Knowledge

To many people, it is implausible that additional people cause more technical knowledge and advance in productivity, ceteris paribus. One source of misunderstanding is the common belief that new technical knowledge usually arises spontaneously, and without connection to social needs. But there now is ample evidence that increased output and investment in a given industry induce more inventions to be made and applied. This "demand-side effect," as economists call it, can be seen in systematic studies of learning by doing, where the time required to complete an airplane or ship decreases as more units are made.

The effect can also be seen in systematic studies of comparative productivity in the industries in the United Kingdom and in Canada that are relatively large and relatively small compared to the same industries in the United States (Simon 1981/1996, chapter 27; Bernal 1953/1970). In the case of electricity, for example, "The barrier, or rather the absence of stimulus to advance, was economic. Electricity developed quickly when it paid, not a moment before" (Bernal 1953, 1970, 131). And a large population size and density imply higher total demand, ceteris paribus, which is why Edison's first street lighting was in New York City rather than in Montana. It is also clear that countries with more people produce more knowledge, assuming income is the same, e.g., the United States as compared with Sweden. And Bernal shows how the power of final demand works indirectly, too. "Once electric distribution on a large scale was proved feasible and immensely profitable, then came a demand for large efficient power sources," leading to the development of turbines (Bernal 1953, 1970, 129). And the development of light bulbs led to advances in creating vacuums, after the subject "had stagnated for about two hundred years. . . . Here was another clear case of the law of supply and demand in the development of science and technology" (Bernal 1953, 1970, 125).

On the "supply side" there is also much misunderstanding, especially in the belief that the number of potential inventors does not matter. One source of this misunderstanding for some is the idea that, to paraphrase, "one need only contrast innovation and creativity in tiny Athens in the Golden Age with monstrous Calcutta" now, or Calcutta with Budapest of the 1930s, to see that more people do not imply the production of more technical knowledge. This argument leaves out the all-things-equal clause; Calcutta is poor. And, underlying this argument is the implied (but unwarranted) assumption that Calcutta is poor *because* it has so many people.

If we make more appropriate comparisons—comparing Greece to itself and Rome to itself during periods with different population sizes and growth rates, and industries of various sizes in different countries now—we find that a larger population is associated with *more* knowledge and productivity, because there are more potential inventors and adopters of new technology. Graphs that plot the numbers of great discoveries, and the population sizes in various centuries in Greece and Rome, bring out this conclusion very nicely (Simon 1990, 200–201).

Improvements in material well-being do not require geniuses. Ordinary people will do. The story of electricity and power production is again illuminating. Bernal describes the "stumbling progress of the first fifty years from 1831 to 1881 . . . the effort put into the development (1831–1881) . . . was small." The people who made the necessary technical developments "were not geniuses . . . and others no more gifted could have hit upon these ideas earlier if the field had attracted enough workers" (Bernal 1953, 1970, 130–1).

Differences in Conceptions of Human Nature

Differences in conceptions of human nature are at the root of much disagreement about economic issues, and evidence about the validity of these different views is relevant to decisions about the economic issues themselves.[1] For example, the doomsayers who desire more government intervention in the production and consumption of natural resources and the optimists who argue for nonintervention of the government in resource markets differ in their views of how individuals and private enterprises behave in the face of economic opportunity; they also differ in their views of the performance of government personnel and agencies entrusted with economic tasks.

I am not suggesting that government should play no role in our economy. But given opportunity, private enterprises will supply more ventures than doomsayers expect, more quickly, and at less cost to the public—especially in the field of natural resources—partly because individuals rather than taxpayers bear the costs of the failing ventures.

Another difference in views of human nature concerns its changeability. Reformers, starting perhaps most vividly with William Godwin (to whose writing Malthus's *Essay on Population* was a response) usually believe that human nature is quite malleable—for example, that self-interested behavior can be rechanneled by the proper social environment. This belief is very important in Marxism; it implies that one can design a social system that has particular desired properties, and then expect people to be molded to fit that system. In contrast, the Scottish moralists—David Hume, Adam Smith, and their teachers and friends—tended to see human nature as relatively immutable, which implies choosing a social and economic system that produces the best results given that fixed human nature.

Society Should Not Be Organized Like a Family

In a family, members share goods out of love and altruism, and their decisions about individual and family activities are

[1]Unfortunately for the discipline of economics, that explicit focus of attention has been lost in the mathematics that constitutes so much of modern "sophisticated" and "rigorous" and "elegant" economics.

(at least sometimes) affected by caring thoughts for one another. But this mode of social organization cannot work nearly as effectively when

- individuals cannot know the preferences of all others in the society;
- the capacity of individuals to empathize with each other is diminished by lack of kinship;
- there is no accepted hierarchy in society as there is between parents and children; and
- the number of goods and possible transactions is very large.

But many persons find it abhorrent to turn over the function of distribution to the impersonal market. And market distribution seems especially abhorrent when the goods seem to have (though they may well not have) a particularly inelastic supply and are especially important to physical survival—for example, food, land, and clean air and water.

Centralized Control of Important Activities

Hayek (1952) thought that the belief in centralized control of economic activity in society is a misplaced analogy to the way engineers plan a dam or bridge, and he traced socialist theory back to the creation of the great engineering schools in France at the turn of the 19th century.[2] Whether or not his account of intellectual history is correct, Hayek's analysis of the contemporary sources of the belief in the need for control surely is sound. Many people believe that without planning and controls, the system just cannot work well. For example, in a debate over whether Champaign County, Illinois, should permit rezoning of farmland for industry, people were heard to say, "I'm for growth, but for *controlled* growth, of course." When you ask them why growth must be controlled by a

[2]This section is heavily influenced by Hayek's works. There also is a fair amount of common ground here with the literature on why people are attracted to socialism (e.g., Kristol 1978; Beckmann 1978; Mises 1972), because resources and environment are part of the "economic problem" that socialism purports to "solve."

planner or an agency, they look at you blankly, as if you are lacking in elementary intelligence.

Many seem to fear that anarchy is the inevitable result of lack of centralized control. Hayek argued that this belief in the need for control is related to a lack of understanding of how a large group of people, acting without any prearrangement, can develop an orderly structure of production and exchange based on individual desires and perceptions of others' desires and intentions. He also mentioned the common failure to understand the difficulty of organizing an economy nearly as well by central planning, even with the aid of unlimited computing capacity and the most detailed information-gathering imaginable, as with a market. These are subtle ideas, not easy to grasp. It is not surprising that even well-educated laypersons often have not thought them through and do not understand them.

Do Ordinary People Require Guidance?

Another possible reason people believe in the need for a centrally directed society is the belief that *others* who are not so well-educated and intelligent cannot figure out how to conduct self-supporting lives that will also thereby contribute to economy and society. The belief that welfare support will be necessary for immigrants—who are often thought (wrongly) to come to the United States with little education and knowledge of English—stems from the arrogance of educated people.

Beckmann and others have suggested that this view fits with intellectuals' desire to be needed by the society, and with their belief that their trained intellects should therefore achieve for them places of special importance and reward in the economy and society. As Beckmann says about a capitalistic society,

> The highly skilled jetliner pilot and the lowly cleaner of sewage systems get a reward beyond dollars—the heady knowledge that they are voluntarily supported because they are genuinely needed. Such a reward is unknown to the professor of Turkish medieval poetry (1978).

In Western civilization this idea is found in Plato. As Popper put it, Plato "charmed all intellectuals with his brilliance, flattering and thrilling them by his demand that the learned should rule" (1966, 199).

Along with this lack of belief in poor people's capacities to run their own lives well is likely to come disbelief that others—and especially the uneducated and poor—can really *create* resources by way of creating new ideas. Perhaps this disbelief is due in large part to a popular lack of understanding of how such human intervention lies behind the resources that we take for granted, e.g., the fertile Midwestern prairie that was a malarial swamp before settlers drained it at great expense in lives and material resources.

Externalities May Not Be All Bad

Environmentalists worry that the unintended by-products—the "externalities"—of humankind's economic activities (especially those that affect the environment) are malign even if the direct effects of production and trade can be benign. But I believe a case can be made that even activities that are not intentionally constructive usually leave a positive legacy to subsequent generations. That is, even the unintended aspects of humans' use of land (and of other raw materials) tend to be profitable for those who come afterward.

Take as an example the "borrow pits" by the sides of turnpikes, from which earth is taken for road building. At first the pits seem a despoliation of nature, a scar upon the land. But borrow pits turn out to be useful for fishing lakes and reservoirs, and the land they are on is likely to be more valuable than if the pits had never been dug.

Another example is garbage disposal. Later generations may find garbage dumps profitable sources of recyclable materials. Even a pumped-out oil well—that is, the empty hole—probably has more value to subsequent generations than does a similar spot without a hole. The hole may be used as a storage place for oil or other fluids, or for some as-yet-unknown purposes. And the casing that is left in the dry well might be reclaimed profitably by future generations.

The explanation of this general phenomenon is that humans' activities tend to increase the order and decrease the randomness of nature. We tend to bring like elements together, to concentrate them. This property can be exploited by subsequent generations. Furthermore, humans perceive order, and create it. One can see this by looking from an airplane for the signs of human habitation. Where there are people (ants, too, of course) there will be straight lines and smooth curves; otherwise, the face of nature is not neat or ordered.

Many acts that we tend to think of as despoiling the land actually bestow increased wealth upon subsequent generations. Of course this proposition is hard to test. But perhaps a mental comparison will help. Ask yourself which areas in central Illinois will seem more valuable to subsequent generations—the places where cities now are, or the places where farmlands are?

One sees evidence of this delayed benefit in the Middle East. For hundreds of years until recently, Turks and Arabs occupied structures originally built by the Romans 2,000 years ago. The ancient buildings saved the latecomers the trouble of doing their own construction. Another example is the use of dressed stones in locations far away from where they were dressed. One finds the lintels of doorways from ancient Palestinian synagogues in contemporary homes in Syria.

Average versus Extreme Values

Safety engineers focus only upon the dangers of a projected line of activity, and urge us to "play it safe." Disputes between those who focus on *aggregate* effects *on average* and those who focus on "worst-case analysis" are common in environmental and natural resource analysis. Much of the thinking of the environmental movement seems to be worst-case analysis.

Nuclear power debates provide many instances of what we might call the safety engineer syndrome. Those who are against nuclear power point to scenarios conceivably leading to, say, 50,000 deaths. Proponents of nuclear power point out that the risk that such a scenario will occur is minuscule, and the "expected number of deaths"—using "expected" in the statistical sense—is very small. The anti-nukes are not impressed by such a probabilistic argument, saying that the worst case has a meaning to us that cannot be treated as part of any set of averages. Nor are anti-nukes impressed by other examples of similarly large worst-case risks that we routinely accept, such as those of power-providing dams that might break and kill hundreds of thousands of people, or of an airplane that might fall from the sky into a stadium seating 70,000 people where all might be killed—risks that are probabilistically greater than those from nuclear energy.

There seems to be a value judgment at the bottom of the argument, a value that cannot be rebutted logically any more than other values can be rebutted logically. But it is possible to point out the costs of such policies that are being neglected in the discussion. It is appropriate for a safety engineer not to be concerned with the costs of avoiding a dangerous activity, because the cost/benefit calculation will be made at higher levels of management. But in any discussion of such activities as nuclear power, it would seem that all discussants have an obligation to have a balanced view and not just focus on one side of the matter, because there is no arbiter in a court of public opinion who will take into account all sides of the matter, as higher levels of management are responsible for doing in an industrial setting.

The case of hydroponic vegetable growing may sharpen the argument. Hydroponics is now a profitable operation around Washington, D.C., for a good many farmers during the months when vegetables are not grown outdoors nearby (Davis 1984). Hydroponic farming takes up only about 1/12th as much land as does ordinary agriculture, the article points out. Shortage of cropland for growing food is one of the common arguments that population growth should slacken now and must eventually cease. But the mention of hydroponic farming usually evokes a long series of what-if objections. What if there will be a shortage of water? Of chemicals? Of sunlight? Of glass to build greenhouses? And on and on. It is impossible to rule out every imaginative scenario without detailed analysis. And of course there is always the seemingly unrebuttable objection:

This cannot go on *forever*. We could even run out of room on earth for hydroponic farming. (Of course there is plenty of room *in space* for spaceships carrying hydroponic farms, a possibility for which the technology is already available without even waiting for further developments. And hydroponic farms can be operated as multistory plants with artificial light.) Each of these questions is offered as an argument against change and growth; the questioner would have us proceed as if hydroponic farming is not a real option.

4. Why Does the Public Not Hear Sound Environmental Thinkers?

Can the amount of false bad news be reduced? Can erroneous beliefs in present and future crises be countered? In fact, the situation *has* improved in the past two decades, because there are now at least some voices and organizations that work to refute false bad news about the environment and social issues.

The public does not hear the truth about environment, population, and various social issues, but the public does hear the doomsaying messages. This chapter argues that the people who know the truth too seldom speak out. As Edmund Burke said, "All that is necessary for the triumph of evil is that good men do nothing."[1]

The Fragmentation of Knowledge

There is an operational difficulty that hinders effective action against false bad news—the practice of fighting fires

[1]Thanks to Kathy Rochelle for providing the exact quote while correcting the typescript. Simon Kuznets was the preeminent population economist of the 20th century, along with being the architect of the national-income system that is now the foundation of the official economic statistics system of the United States. He was the greatest student of economic-demographic history and economic development, and a Nobel prize-winner. When Kuznets addressed the issue of population growth in the 1950s and 1960s, the prevailing professional thought was pessimistic. "My impression is that recent professional (and popular) literature has emphasized the disadvantages and dangers of population growth" (1965, 124). And even with all of Kuznets's prestige, he worried that he would be considered ridiculous and discredited if he emphasized the positive aspects. "The concluding remarks are addressed primarily to qualifications, to avoid dismissal of this discussion as an expression of exuberant but unfounded optimism" (1965, 137). He did not address the general public on the matter in the 1970s even when no-growth ideas were being trumpeted that were entirely at variance with the facts as he knew them. When I asked him whether he ever felt the desire to make known the facts, he replied, "If I did that, I would not be able to get my work done." How should one think about Kuznets's decision?

of untruth only when the fires flame up. For example, when the Rio Earth Summit meeting came along in 1992, a variety of organizations in and out of Washington, D.C., began scurrying to cobble together responses. The Cato Institute prepared a list of scientists who would present the evidence rebutting false scares and whose names it gave to the press. The Competitive Enterprise Institute worked up some source materials in conjunction with the National Center for Policy Analysis in Dallas and the Political Economy Research Center in Bozeman, Montana. And there were other initiatives. But these efforts were too little, too late, and most especially too fragmented to even slow the juggernaut of the environmental movement on that occasion or in general.

The fragmentation of knowledge and lack of integration of effort result in academics' accepting falsehoods outside their area of expertise (Lindzen 1992). Edward Krug (1992), who was a key figure in the acid-rain fiasco, observed that he accepted the environmental propaganda that acid raining on the above-ground parts of trees was damaging the trees. But, as a scientist, he could find no evidence that acid rain affects soils, my area of expertise. Then he found out that the forest scientists—their specialty being the above-ground parts of trees—knew that acid rain was not harming the above-ground parts of the trees. However, they accepted the environmental propaganda that acid rain was harming trees from below the ground, that is, through soils.

Consider also the case of Haroun Tazieff, an eminent volcanologist who served as French secretary of state for the prevention of natural and industrial disasters. Dioxin was the first issue into which he inquired when appointed to that post.

> Here are two examples of articles that prompted my investigation of the subject: "Seveso: The Hiroshima of Chemistry," and "Seveso: 9 Months After: The Lessons Were In Vain." Both appeared in the popular science periodical *Que Choisir* [What Choice] in April 1977.
>
> I was at the time a Citizen Lambda [John Doe], an individual among hundreds of millions targeted by the disinformation campaign launched on a global scale. I had believed

> in what was thus universally and imperatively affirmed as incontestable truth: that PCBs, and the dioxins they emit when heated to 300° Celsius, were frightful poisons. One or two years of this propaganda had led government officials—just as incompetent as I was in matters of polychlorobiphenyls—to make them officially illegal.
>
> A half-dozen years later, I found myself responsible for the prevention of disasters, natural and technological, for the French government. The natural ones I knew quite well, since they are related to my profession. As for technological disasters, it was necessary to inform myself. The very first dossier I asked to have delivered to me—so much had I been convinced of the extreme hazard of PCBs—was the one on the explosion at the chemical plant in Seveso, Italy, in July 1976. The study of this dossier and the inquest I led at the time revealed to me, first of all, that this so-called catastrophe had not one single victim. (This gives the "Hiroshima of Chemistry," as it had been baptized by an ostensibly serious monthly science magazine, a tinge of anticlimax.)
>
> Second, I learned that dioxins, according to the judgment of all the actual experts consulted (and of the very knowledgeable Academy of Science), are not at all "frightful" and have never, anywhere, killed anyone. . . .
>
> Presenting the industrial accident at the ICMESA factory in Seveso as an apocalyptic catastrophe was a matter of deliberate disinformation—in less diplomatic language what one calls a lie (Maduro and Schauerhammer 1992, viii, ix).

Lindzen, Krug, and Tazieff are rare in spotting false scares outside their own special fields. In 1983, geographer John Fraser Hart of the University of Minnesota sat a whole day with a dozen other contributors to the volume *The Resourceful Earth* that Herman Kahn and I were organizing and editing. Hart then observed that everybody in the room was optimistic about his own subject, but pessimistic about everybody else's subject. And it was true, all agreed. (But why should that be?) Only Herman and I were across-the-board optimistic, and that was because our work had caused us to touch on the entire range of topics.

This phenomenon is apparent everywhere. Physicians know about the extraordinary progress in medicine that they fully expect to continue, but they can't believe in the same sort of progress in natural resources. Geologists know about the progress in natural resources that pushes down their prices, but they worry about food.

Even worse, some of those who are most optimistic about their own areas point with alarm to other issues to promote their own initiatives. The motive is sometimes self-interest. Some examples:

- The scientists of the Space Exploration Initiative at Los Alamos are excited and confident about the challenges they themselves address. But they justify their own adventures as intended to "relieve the stress on Earth's environment from population growth, and provide our world with limitless resources for the future," because "as Earth's assets dwindle, Mars may offer resources mankind will need" (Synthesis Group 1991).
- Noel Davis, who runs PhytoFarm in DeKalb, Illinois, and produces a ton of food every day in his hydroponics factory on *one acre of land* (yes, you read correctly)—enough to feed 500 or 1,000 people—a process that makes land almost irrelevant as a factor of production for food—justifies his operation on the grounds that we are running out of farmland. "Each year the United States is losing an area of farmland greater in size than the state of Rhode Island," he writes (Field 1988, 51).
- An inventor with a remarkable system for making waste into products of value while reducing pollution from the system almost to zero—Leonard (and Frank) Keller of Methacoal—feels the need to make worrying counterfactual statements about running out of landfill space.

I have a special interest in all this, because population economics is the toughest nut of all. Even those who otherwise are optimists worry about it. For example, H. W. Lewis's thoughtful, informed book *Technological Risk* discusses a variety of environmental worries, providing data and careful analysis at every juncture. But about population growth he offers nothing more than a flat assertion that it is a grave danger, without facts or analysis or reference to any scientific literature:

He begins: "The prevention of nuclear war [is] second only to overpopulation as a real and immediate threat to the human race" (p. 284). And he writes in his introduction: "The time scale for solving the population problem is one or two generations. It cannot go on this way, and the die is already cast" (p. xiii). Yet his entire book is an effort to subject such loose thinking to close analysis.

A group of well-known scientists who issued the scientifically accurate and generally optimistic assessment of environmental issues entitled The Heidelberg Declaration took a big step in the right direction. But even there one finds that they express worry about overpopulation, though without any documentation. Population is the ultimate taboo, it seems.

Environmentalists Speak with One Voice

In contrast, the anti-growth environmental movement speaks almost with a single voice. Many organizations have banded together in the *Global Tomorrow Coalition*. And when their foremost spokespersons—for example, noted biologists Peter Raven of the Missouri Botanical Garden, and Paul Ehrlich of Stanford—give speeches, they address the entire litany of worries, crying that each and every issue is a threat and part of a general crisis. And they cite surprised scientific consensus. They bludgeon us with "700 Members of the National Academy," "26 Nobelists," and the like.

The power of the environmental and related organizations to muster a strike force on any issue is awesome. A full-page advertisement in the *Washington Post* and probably elsewhere was headed: "SABOTAGE! of America's Health, Food Safety and Environmental Laws." It turned out to be about the GATT trade rules, and the sponsoring organizations included the Citizens Trade Watch Campaign and its parent, Public Citizen, along with Sierra Club, American SPCA, Friends of the Earth, and many others from the environmental organization crowd. Clearly this was not a true environmental issue, despite the headline and the photo of dolphins, but the scratch-my-back philosophy is operative to an extraordinary degree.

In contrast, the careful-thinking anti-scare scientists invariably restrict themselves to their own subjects, for two reasons: First, as careful researchers they limit their statements to what they know expertly—in many ways, an understandable and admirable personal policy. Second, about subjects they have not studied, they know only what all the rest of the public knows—that which is derived from the newspapers and television, and which expresses the doom scenario. So they are pessimistic about these other issues, as Fraser Hart noted.

Needed: An Organization Dedicated to the Entire Truth

The world needs an organization to be a Truth Lobby, whose mission it is to combat *all* false information when it is disseminated—for example, phony Alar scares, false assertions about acid rain, false statements that immigration is at a historical peak, and false statements that DDT causes more harm than good. Issues such as global warming would be tough for a Truth Lobby because in such a case the facts are not easy to come by in a short period of time after the public hears a new alarm. But inability to respond to every issue that *some* would like to respond to should be a strength rather than a weakness.

Each of the existing organizations that work for growth and freedom, and against the doom scares, is organized for special limited purposes. Several think tanks have gotten interested in particular aspects of the environment and resources issue. As of 1997, a few trade organizations, such as the Western Coal Association, have been willing to confront the anti-growthers broadly. But most, such as the nuclear industry's USCEA and the Chemical Manufacturers Association, tend strictly to their own knitting—and then wonder why nobody comes to their assistance (although through its aid to Elizabeth Whelan's American Council on Science and Health, the CMA does strike a wider range of blows). It is imperative that these trade associations come to see that the private welfare of each is best served by joining forces with others who are working for truth in related domains. The motto for them should be: Ally or die.

The country is also dotted with isolated workers for the truth about environment and resources, some of whom have nothing in common with each other except this complex of issues.

- Bill Stonebarger's Hawkhill Associates in Madison, Wisconsin, sells audio-visual materials by mail to school science programs and goes beyond the clichés to present the facts about such matters as the safety of nuclear power and running out of energy.
- I. W. Tucker's tiny National Council for Environmental Balance runs a small mail-order bookstore publishing and retailing little-known truth-telling books such as those of Dixey Lee Ray.
- Andrea Rich's Laissez Faire Books sells libertarian tomes as its main mission—including some books that will be anathema to some conservatives and liberals—but this mission includes selling lots of books that show how free markets can lead to enhanced environments.
- S. Fred Singer's Science and Environmental Policy Project brings the true light of science to many issues.
- Greg Rehmke runs a program to provide support for high school debate programs, and the subjects frequently are environmental and population matters.
- Elizabeth Sobo is a one-woman investigative newsroom devoted to truth about population, and can come under this umbrella, though her general economic view will flip out any libertarian.
- Max More's *Extropy* magazine focuses on apocalyptic environmentalism.

But all these missions, and the many people who protest against environmental falsity by writing letters to their local newspapers, have no organizational means of connection.

If defenders of the truth are to move beyond scattered and ineffective responses to the doom establishment—which now add up to little more than lying down in front of advancing tanks—the separate activities and ideas must come together with an organized message. The first step should be for the think tanks to inform themselves, and the scientists who individually have anti-doom messages, about the entire spectrum of issues. They must enable themselves to recognize that there is a general tendency for all things involving human welfare to be getting better.

Indeed, the message about human betterment and economic progress is more general than any individual statements about raw materials, air, water, life expectancy, education, and the like. There is solid theoretical basis for the idea that *all* aspects of human welfare should get better, not just as a matter of coincidence but as part of a broad causal mechanism. Humanity has necessarily evolved so that we have more of the nature of creators than of destroyers—or else the species would have died out long ago. People seek to improve their conditions, and therefore on balance people build more than they tear down and produce more than they consume. Hence each generation leaves the world a bit better in most respects than it begins with.[2]

After individual scientists become educated about aspects of the situation beyond their own topics and are brought on board the good vessel Progress, the separate organizations can then create an umbrella organization, though without diluting their own special interests. And then they can sing in chorus when it is necessary, rather than singing individual solos one after another.

Of course, a responsible organization will accommodate itself fully to the desire of scientists not to be associated with any statement that they cannot feel completely comfortable with, and no compromises. This means that no one's name should appear without prior approval. And it means that statements should be limited to matters of fact and not of policy, as a general rule.

One of the most important functions an umbrella organization can perform is to put a card into journalists' Rolodexes. Reporters assume that if there is a point of view on a subject,

[2]Of course there are exceptions to progress, such as whether the health of old people now is systematically getting better decade after decade or not. But certainly the broad generalization holds, and not just for resource and physical environment issues, but for all other issues pertaining to the standard of living.

there must be an organization to represent that point of view. For example, following an interview by a BBC crew with Robert Whelan that was omitted from a broadcast on population growth, the chairman of the BBC defended the action as follows: "The editor of 'Nature' tells me it is difficult to find academic organizations supporting your arguments" (Marmaduke Hussey to Whelan, Dec. 10, 1990). Absent an organizational presence, they assume that no body of information or opinion exists. There is a lacuna here that desperately needs to be filled.

The same is true at the grassroots level. When my wife and I went out for a bird walk recently, we found ourselves at an Audubon Fair. Represented at the fair were solar electricity sellers, a clean water organization, Zero Population Growth, and many more such. But there was not one sign of the free-market environmentalists or the anti-scare groups. The same is true of every Earth Day demonstration. It is not surprising that children grow up believing the environmental activists they hear on television, and read about in their textbooks, because they never find out that there is any other side to the issues.

5. Why Are So Many Biologists Alarmed?

All hail to biologists! By greatly improving our agricultural productivity and increasing our understanding of the processes that affect sickness and health, biologists have contributed far more to human progress and contemporary well-being than have members of any other discipline. We owe them our gratitude.

We do not, however, owe biologists attention when they talk pure nonsense about subjects entirely outside their field of special knowledge.

Certainly in the past few decades—and perhaps as far back as Thomas Malthus's and Benjamin Franklin's time—the most strident prophets of doom about environment, resources, and population have been biologists, with some handmaidens among the physicists and chemists. The bestsellers among the book-length warnings of disaster due to population growth have been William Vogt's *Road to Survival* (1948), Fairfield Osborn's *Limits of the Earth* (1953), Karl Sax's *Standing Room Only* (1960), and Paul Ehrlich's *Population Bomb* (1968), all by biologists, and there are literally scores more books by biologists that are lesser known; *The Limits to Growth*, produced by MIT (Meadows et al. 1972), is a rare exception in this genre.

Biologists have consistently warned about supposedly impending shortages of the staples of economic life—food, farmland, raw materials, a clean environment, species of wildlife—to the peril of our standard of living. And they have asserted that the quality of human life has gotten worse rather than better. They even threaten us with economic doom, a subject they claim to speak about with more authority than economists.

Biologist Mary E. Clark denies that "those who have been trained in modern 'economics' actually deal with economic realities" (Clark 1989). Another well-known biologist asserts, "Perhaps the most serious single academic problem in the world is the training of economists" (Raven 1988, 229). One reads that "economic theory was developed at a time when human population was small and the planet was considered an infinite resource" in *The ZPG [Zero Population Growth] Reporter* (September 1991, p. 1). And as Ehrlich put it (please forgive the personal examples to come), "Economists confuse 'pollution' with the more serious problems of loss of ecosystem services," referring to "the sort of blunders Simon and other economists of his like commit when they attempt to deal with problems of population, resources, and environment." And two prominent environmental activists combine to make this statement:

> Another impediment to perceiving and acting on overpopulation has been conventional economics, which activist and author Hazel Henderson has called "a form of brain damage." . . . For example, economist Julian Simon, whose work has served as a rationalization for recent U.S. population non-policy, maintains that population growth generates its own solutions" (Mills 1991, p. 48).

Why do biologists behave so? Three personal anecdotes introduce the more general discussion. At a social occasion, a very, very distinguished biologist—let him remain nameless except to say that he heads a major scientific institution in the Washington area—confidently and vigorously made a collection of distressing assertions about the economic condition of graduate students. When it was suggested that his judgments did not square with aggregate statistics, he dismissed census data and historical trend data as irrelevant, and inferior to the personal observations and data he is familiar with at a single university, and to the journalistic reports that he had read in the newspapers.

When I debated distinguished ecologist Garrett Hardin about whether an increase in population has good or bad effects in the long run, the large audience was mostly composed of biologists and their colleagues in "environmental studies." Hardin was openly contemptuous of all data, but especially statistical analyses of such phenomena as immigration and unemployment, asserting that the Malthusian theory was more to be believed than empirical refutations of it. "One cannot expect much in the way of secure truth from statistics," he said. Instead, we should "use such theory as we have that looks secure, that makes sense, and see if we can't make sense of the world using this theory." The audience applauded Hardin's remarks, and not a single critical question was asked of him in connection with his rejection of historical experience.

Bernard Davis, a distinguished member of the Bacterial Physiology Unit at Harvard Medical School, who speaks his mind unusually freely, wrote to me, "I am convinced *as a biologist* that in the long run the problem of population growth is likely to turn into the world's most oppressing problem" (letter of October 26, 1989, italics added). No data were cited.

The organizations most influential in our national life on these matters, and on such related matters as (human) population growth and in-migration, have been biologically oriented, from A to Z—the Audubon Society to World Wildlife Fund to Zero Population Growth. And in the quotation above, Davis implies that he has a special point of view *as a biologist* about a matter whose truths one would think are independent of a person's occupation. Would a scholar make assertions about rent control or cancer research or population growth "as a chemist" or "as a Chinese philologist"?

Over the years I have wondered about the modes of thought involved in the thinking of people at large that lead them to make consistently erroneous judgments and prophecies about materials and environment. Other chapters of this book have identified a long list of causes of the wide dissemination of judgments that do not have scientific foundation—for example, the penchant of the newspapers for bad news. But the specific characteristics of the thinking of *biologists*—I speak about the category for convenience, though of course there are important exceptions—have not previously been discussed, to my knowledge. And in light of the importance of biologists in this movement, the topic calls for special treatment. The biologists will be quick to point out the limitations of an outsider in analyzing the thinking of biologists. But two decades of biologist-watching in the context of doomsday scares provides some useful experience.

In my view four characteristics of biological research lead to incorrect assertions about natural resources and the environment:

- the nonhistorical character of most biological work;
- the experimentation with individuals rather than samples of a population;
- the study of adaptation in nonhuman animals; and
- the inappropriate use of biological ideas in other settings.

I will consider each of these cognitive elements in turn.

Biology Is Largely Ahistorical

Except for the study of evolution, biology is an ahistorical field of inquiry. There is little reason to think that eagles or human kidneys are very different now from what they were seven millennia ago. And individual "human nature" may be much the same now as then. But society's powers to produce goods and to develop new capacities are vastly greater than earlier; consider the difference between the technological responses to the Black Death and to AIDS.

Applying the assumption of constancy in society's responses to a food or fuel shortage, or to a change in climate, since (say) 5000 B.C. until now, leads to grave [sic] mistakes. Malthusian famine, for example, no longer occurs on this earth—witness the overflowing shipments of food to starving countries in Africa. The causes of starvation no longer include the physical inability of society to transfer food to starving people, or the incapacity of cultivators to plant, harvest, and store enough for their needs. To continue to think about starvation as it was in earlier centuries leads to unsound policy prescriptions.

It is hard to square a reasonable perspective on human history with this sort of statement:

> If you asked me right now, "Has science been a net benefit or a net harm to mankind?" I have to say "I don't know." So far it's done about as much harm as it's done good. Maybe a little more harm than good. Americans are part of the lucky group, but most people in the world are living considerably more miserable lives than they lived six or seven thousand years ago. We are the most vulnerable population the world has ever seen. We're more crowded. We have more undernourished people than have ever lived before (Paul Ehrlich, quoted in McBride 1980).

Inexperience with phenomena that change over time may also be responsible for biologists' looking at too short a series of experiences. An example is taking a single hot summer as the harbinger of global warming. Another recent and costly example was the conclusion, based on just a few years of price rises about a decade ago, that petroleum in particular and energy in general are becoming more scarce. (Remember biologist Paul Ehrlich's "What will we do when the pumps run dry?")

Sometimes centuries are required to determine whether a trend is occurring. Observations covering a period even as long as 50 years in the history of the Nile River—where a drop in the river's level can continue for that long a period—can make it look as if the Nile is going dry. Only data covering many centuries put half a century into proper perspective (Baumol and Oates 1995) and prove that one can draw the correct conclusion—that in the long run the Nile is neither rising nor falling very much—only if one gathers and observes the data for at least that long a period. The same is true with trends in climate; one can easily be fooled by looking at temperatures for just a year or a decade (or even a century). Biologists' work does not equip them with experience to deal with such long time-series.

Biologists Study Individuals Rather than Samples of Populations

Though variability in results can be important in biological experiments, variability is less central a phenomenon than in social science. If penicillin kills bacterial infection in one case, repeated study in the laboratory can refine the conditions under which the effect will occur reliably. Jonas Salk and his coworkers were able to draw tentative conclusions from a test of an AIDS vaccine in three monkeys, one of which did not have the virus and two of which did have it.

In contrast, the connection between, say, population density and economic growth is "only" statistical. No matter how you look at a collection of observable entities such as countries, there will be many glaring exceptions to the main tendency, and the main tendency is likely to hold with a correlation that is closer to randomness than to perfect relationship. Looking only at particular instances—e.g., at a particular country that is far from the pattern—prevents drawing sound generalizations.[1]

Related to the idea of variability is the idea of representativeness. Biologists usually do not work with the idea of a representative sample from the universe of interest, and they do not need to understand the properties of biased and unbiased samples. Therefore, they are not protected against error from grossly unrepresentative samples. For example, one of the estimates of the rate of tropical deforestation most relied upon by biologists who forecast the denuding of tropical forests comes from Sommer's (1976) study that drew observations only from the *sides of roads* through the Amazon and elsewhere. Sommer then simply assumed that the forests would be similarly cleared for four kilometers on each side of the road, and then projected the number of kilometers of roads that would be built. No matter how clever the theoretical assumptions one makes to supplement these observations, such evidence

[1]There is a long history in medicine of disdaining statistical evidence in favor of "clinical" learning—the rejection of Ignaz Semmelweiss's discovery of the sources of childbed fever being a leading horrible example. The preference for conclusions drawn from the physician's own "clinical" experience rather than statistical reasoning has become almost ludicrous at a time when the main basis of judging the efficacy of drugs and operative procedures is the controlled experiment, analyzed with statistical techniques, published in technical journals that are persnickety about the use of such evidential methods.

is most unlikely to help one soundly judge the state of the forests that are *not* near the roads—which are the bulk of forests, of course.

The distrust of aggregate statistics by many biologists may derive from their relationship to representative individuals rather than samples. I wish I had a dollar for each biologist who, when I show aggregate trends toward better nutrition in the world over the decades, and decreasing scarcity of raw materials over the centuries, has recited—just as Garrett Hardin did in our debate—the old saw about lies, damn lies, and statistics. And they ask, "How do we know that *your* statistics are valid?" before even inquiring into the provenance and nature of the statistics.

An example of this is the comment of a specialist in forensic medicine at a drug conference in Israel: "My one advantage over the other speakers is that I have seen firsthand the damage drugs do to the lungs, liver and kidneys of addicts," he said.[2] His own visual observations, and his emotional reactions to them, made him feel qualified to render a policy judgment without considering the many other aspects of prosecuting drug users and sellers. This is an example of the fallacy that psychologists have demonstrated to occur when vivid evidence is given more weight than it deserves, relative to pallid evidence such as data (Nisbet and Ross 1980, chapter 3).

Animal and Human Adaptation Differ

The subject of biological study usually is either a nonhuman organism or is at a level of organization below the organism. Concepts that are appropriate for nonhuman organisms—niche, carrying capacity, etcetera—are inappropriate for the creative aspect of human beings that is the central element in long-run economic activity. Biologists then consider the human to be "just" an animal. As Hardin put it, "Since this is true for all other populations, it's hard to see how it can be false for the human population unless you say we're just utterly different from other animals, which a biologist is not

[2]*Jerusalem Post*, April 29, 1989, p. 19.

willing to do." We may be animals, but we are not *only* animals: not all that is important about us is also true of other animals. The 1960s and 1970s generalization from the work of Calhoun on Norwegian rats to policy recommendations for human society has been a classic example of this muddle.[3]

It may be that working with nonhuman species predisposes a person to have relatively little faith in the adjustment capacities of human beings. For example, in regard to the possibility of fusion energy, Paul Ehrlich was quoted as saying that cheap, inexhaustible power from fusion is "like giving a machine gun to an idiot child."[4] Hardin says, "People often ask me, well don't you have faith in anything? And I always have the same answer, I do have one unshakable [faith], and that is I have an unshakable [faith] in the unreliability of man. I know that no matter what we do, some damn fool will make a mess of it."

Inappropriate Use of Biological Concepts

Many biologists make astonishingly brave assertions about subjects that are wholly outside their fields of research, assertions intended to be understood as expertise rather than as mere lay opinion. Perhaps the amazing success of biological science in recent decades induces such extraordinary confidence in biologists who have not done research on the topic that they will hazard grand statements about, for example, resource economics—without even studying the body of work of economists for whose field the topic has been central throughout its history.

Some biologists go so far as to tell economists that their discipline is fundamentally wrong and they do not understand their own subject matter. Ehrlich, for example, frequently asserts that only an ecological viewpoint—that is, his own viewpoint—can make sense of economic phenomena. Here it

[3]For a discussion of Calhoun's studies in the context of Malthusian thinking, see chapter 24 of Simon (1996).

[4]*Los Angeles Times*, April 19, 1989, Part 5, pp. 1–2.

is necessary to quote at some length to give the full flavor of this criticism:

> It has long been clear to ecologists that the extreme growth orientation of mainstream economics is a major reason that politicians, businessmen, and others advised by economists, as well as the public at large, fail to recognize the increasing seriousness of the population crisis in particular, and the deepening predicament of Homo sapiens in general. Most people do not recognize that, at least in rich nations, economic growth is the disease, not the cure (Ehrlich and Ehrlich 1990, 175).
>
> The failure of conventional economics to contribute to a resolution of the human predicament is understandable from a cursory examination of what economists are taught.
>
> Since they are unaware of the stress that natural systems are now under, most economists believe that the scale of economic activity can be increased indefinitely (p. 176).
>
> The absurdity of the idea of perpetual economic growth . . . (p. 177).
>
> Barnett and Morse had the laws of physics exactly backwards. . . . But even ignorance of physics is not sufficient excuse for the faith of economists in infinite substitutability (p. 180).
>
> In our opinion, whether humanity will be able to move toward a population size and an economic system sustainable largely on income will depend in no small degree on economists (p. 181).
>
> If we are to escape our current predicament, [economics taught from an ecological point of view] should become a major area of specialized education, and replace neoclassical economics as the central focus of economics departments.
>
> Considerable instruction on the basics of how the physical-biological world works must be included in the training of all economists. Otherwise they will continue to whisper the wrong messages into the ears of politicians and businessmen (p. 182).
>
> Economists think that the whole world is just a market system, and that free goods are infinitely supplied. They are a discipline built on transparent mistakes, from the point of view of a physicist or a biologist.
>
> Economists are probably the most dangerous single profession on earth, because they are listened to. They continue

> to whisper in the ears of politicians all kinds of nonsense. Everybody feels that the economic system is what dominates human affairs, when actually the economic system is hopelessly embedded in the physical and environmental systems (Ehrlich quoted in McBride 1980).

The rejection of the economics discipline by biologists often goes hand in hand with the suggested substitution of an energy standard for the basic economic concepts of value theory. Such biologists ascribe economists' analyses to ignorance of physical [sic] science. As Hardin put it, "There is no use beating your head against the wall trying to discover an escape from the laws of thermo-dynamics. . . . Nobody looks for perpetual motion machines now, and only some economists believe in them." But when making this objection, as well as others, biologists seldom refer to economic texts or show any other signs of having studied economic theory and empirical findings.

Perhaps it is not fair to tar all biologists with the brush of the ideas expressed by a few such as Ehrlich and Hardin. But Paul Ehrlich can count as his collaborators, allies, and supporters a very large number of distinguished biologists. He has received such prestigious awards for his work as a MacArthur Prize Fellowship and the Crafoord Prize in Population Biology and the Conservation of Biological Diversity. He was also awarded the first AAAS/Scientific American Prize for Science in the Service of Humanity. To my knowledge, no distinguished biologist has disavowed his statements as expressing the views of the biological profession at large. And his sort of viewpoint is expressed from positions of considerable influence. For example, Robert Goodland, the World Bank's principal ecologist, said that "the most important thing for the environmental movement is to revamp economic thinking."[5] And on the basis of this viewpoint, "World Bank president Barber Conable . . . announced a major reorganization of the bank that will include much more attention to resource

[5]*Science*, May 15, 1987, p. 769.

conservation and the environmental aspects of development projects."[6] This was despite the fact that the principal economists on the Bank's staff did not agree about the proposed policies—and the bank is supposed to be an economic institution with economic justification.

There is an interesting continuity from generation to generation in biologists' views about human ecology. This is Malthus's description of Benjamin Franklin's analysis:

> It is observed by Dr. Franklin, that there is no bound to the prolific nature of plants or animals, but what is made by their crowding and interfering with each others' means of sustenance.—This is incontrovertibly true.—In plants and animals the view of the subject is simple. They are all impelled by a powerful instinct to the increase of their species and this instinct is interrupted by no reasoning or doubts about providing for their offspring—the superabundant effects are repressed afterwards by want of room and nourishment—and among animals, by their becoming the prey of each other (Malthus 1803, 203).

Contemporary biologists, too, offer animal-ecology experiments as analogies to human population growth. Their models include Calhoun's famous Norwegian rats in a pen, hypothetical flies in a bottle or germs in a bucket, and meadow mice or cotton rats, which will indeed keep multiplying until they die for lack of sustenance. Price, in *The 99th Hour*, gives a typical example of this view.

> Assume there are two germs in the bottom of a bucket, and they double in number every hour. (If the reader does not wish to assume that it takes two germs to reproduce, he may start with one germ, one hour earlier.) If it takes one hundred hours for the bucket to be full of germs, at what point is the bucket one-half full of germs? A moment's thought will show that after ninety-nine hours the bucket is only half full. The title of this volume is not intended to imply that the United States is half full of people but to emphasize that it is possible to have "plenty of space left" and still be precariously near the upper limit (Price 1967, 4).

[6]Ibid.

But we must recognize what Malthus came to recognize. After he published the short simplistic theory in the first edition of his *Essay on Population*, and after he had the time and inclination to consider the facts as well as the theory, he concluded that human beings are very different from flies or rats. When faced with the limits of a bottlelike situation, people can alter their behavior so as to accommodate that limit. Unlike plants and animals, people are capable of foresight and may abstain from having children from "fear of misery." That is, people can choose a level of fertility that fits the resources that will be available. And people can alter the limit—expand the "bottle"—by consciously increasing the resources available. As Malthus put it, "Impelled to the increase of his species by an equally powerful instinct, reason interrupts his career, and asks him whether he may not bring beings into the world, for whom he cannot provide the means of support."

Malthus came to stress the difference between the breeding of animals and of humans, and he decisively rejected Benjamin Franklin's animal analogy:

> The effects of this [preventive] check on man are more complicated. . . . The preventive check is peculiar to man, and arises from that distinctive superiority in his reasoning faculties, which enables him to calculate distant consequences (Malthus 1803, 3, 9).

That is, human beings are different from animals in that we have much more capacity to alter our behavior—including our fertility—to meet the demands of our environment.

The biologists whose predictions about resources have been wrong across the board for the last two decades—Paul Ehrlich and Garrett Hardin chief among them—have lost no public credibility and are quoted with as much avidity as ever. No biologist has ever told me of a radical change in his views as a result of exposure to new scientific knowledge on these subjects, not even the 1986 report of the National Academy of Sciences on population growth and economic development. Nor have I heard of radical conversions among biologists from exposure to two decades (indeed, two centuries) of data that contradict the conventional wisdom, though I have heard of such radical changes among every other class of intellectual worker.

This is what is particularly hard to understand about the biologists—their imperviousness to new information on these subjects, and their continuation with the same unrevised views. Garrett Hardin asks, "What is the theory involved in this area?" and answers, "the essentials of Malthus" (Hardin-Simon debate, p. 2). No matter that that theory has not fitted the facts since Malthus, or that there is newer theory that does fit the facts. And even though he notes that history since then has been "rather embarrassing for [Malthus's] theory," he concludes that "basically the Malthusian theory still explains the facts" (Hardin-Simon debate, p. 4). Calhoun's analysis is identical to the assessment of Franklin, and Malthus's crushing retort to Franklin would seem as fresh today as it was then.

William Petersen (1976, 1977) suggested that biologists are unwilling to recognize that social science constitutes a body of knowledge that is not immediately available to the layman. This may help explain their willingness to make sweeping statements about economic and social phenomena without scholarly research. It may also explain the willingness of many biologists to use unscientific language when discussing these phenomena, e.g., likening population growth to a bomb, cancer, a swarm of maggots, and other malign phenomena, as psychiatrist Frederic Wertham pointed out some decades ago (1969).

One might wonder: Are Ehrlich, Hardin, et al. outliers? Certainly there are some, perhaps many, biologists who do not support their views, or do not support them fully. But one does not find similar prominent cases in other fields. Furthermore, biologists who espouse the kind of thinking discussed in this chapter are almost never disavowed or denounced by associations of their colleagues. And the large scientific organization that has been most sympathetic to their views has been the American Association for the Advancement of Science, whose conferences and official journal, *Science*, have been heavily dominated by biology. So it would seem that these most vocal biologists are not unrepresentative of their field.

At Last I Understand the Ecologists

For decades, ecologists have been arguing with economists. Curiously, the ecologists share much in their basic intellectual outlook with economists. Yet they reach radically different conclusions about the environment. This section delves into this particular aspect of the biology-economics divide.

Both groups feel misunderstood, but think they understand the other. Now I recognize that I, at least, have not understood the ecologists well enough, because I have not grasped the grand vision of nature and humankind that underlies their thinking. If I (on behalf of other nonecologists) try to state that vision and why I do not share it, perhaps it will help us talk to each other.

I will try to describe the ecologists' vision by analogy. I came within two weeks of going to medical school after I got out of the Navy in 1956. One reason that I didn't was a preference of mine and others not to take medical drugs except when the need is overwhelming. I feared that my preference would put me at odds with the profession I would be studying and practicing. I'm not sure how much of my preference was due to worry about side effects, and how much to some belief (whose logic I could not pin down) that if disturbance can be avoided it makes sense not to disturb the complex system which is the body. Reading about psychological experiments showing how rats and babies can, under many conditions, choose diets wisely had impressed me, as had Walter Cannon's notion of "the wisdom of the body." I thought it likely that tampering with a very complex system about which we understand so little is inherently dangerous. Medical practice nowadays is much closer to my earlier feelings, of course.

The ecologists' conception of man and nature resembles that conception of the body. Ecologists believe that we ought not make changes in the existing order unless the reason is pressing and unless our knowledge of the likely consequences is extensive, because so many undesirable effects may be indirectly caused by a particular alteration. In *Silent Spring,* Rachel Carson expressed this point of view particularly well with respect to modern fertilizers and pesticides (1962, pp. 17, 20).

Complexity in the system, and insufficient knowledge of it, cause ecologists to worry (if I understand them correctly) that an alteration could knock the whole system out of whack, which will then induce a series of additional changes to compensate, which will then have even worse effects, and so on, until the whole system comes apart or explodes.

> Only within the moment of time represented by the present century has one species—man—acquired significant power to alter the nature of his world (Carson 1962, 16).
>
> To adjust to these chemicals would require time on the scale that is nature's; it would require not merely the years of a man's life but the life of generations (p. 17).
>
> The whole process of spraying seems caught up in an endless spiral (p. 18).
>
> Future generations are unlikely to condone our lack of prudent concern for the integrity of the natural world that supports all life (p. 22).

And as her head quote, Carson gives us this from Albert Schweitzer:

> Man has lost the capacity to foresee and to forestall. He will end by destroying the earth (p. vii).

As I see it, however, this vision—together with the above analogy of the human body and drugs—is not appropriate for understanding and dealing with modern man's economic and environmental situation. Yes, it is possible to avoid large "unnatural" alterations in the way we treat our bodies, in comparison to the way people lived in earlier millennia. Though we cannot have a similar mental environment, we can sleep the same way, exercise in not dissimilar ways, and even eat much the same way if we choose to do so; even the casual diet of a person who is not faddishly given to one kind of food or another probably does not depart fundamentally from the diet of our long-dead ancestors.

But with respect to our macrorelationships to the environment, the situation is entirely different. We cannot say, "Stop the world; I want to get off," or even, "Stop the world; I want to stay on." There is no way we can avoid large "unnatural" humanmade alterations, because our whole life system is composed of manmade artifacts. What happens to our heating, our cooking, our mobility, if we turn off our electric motors, or stop digging in the earth? Nor can we even say, "Things seem to be more or less okay as they are now, so let's keep things pretty much as they are," even aside from the fact that things are not okay in the view of most inhabitants of the world. Furthermore, we can't just go on doing what we've been doing without change, because the existing system—like all existing systems—necessarily runs itself down unless we replenish and refurbish it. For example, a "soft path" to energy and growth such as advocated by Amory Lovins is a charming idea, but it is not likely to be a meaningful option.

Compare (1) our present global commitment to create for all humans a decent sustenance and opportunities for participation in modern society with (2) persons in the year 1900 who might have committed themselves to traveling to the moon. If we satisfy ourselves with simple Wright-brothers technology, we won't make it and instead will probably get killed in a crash soon. If we continue to improve the technology, we take chances of as-yet-unknown (and perhaps more disastrous) hazards than are possible with a primitive airplane. But with time and thought we are more likely to make the system increasingly safe—as has indeed happened with air travel. It is now safer to go to the moon than it was for the Wright brothers to fly 1,000 yards. And with a little luck and a lot of human effort and ingenuity, in the future we may be able to provide a high level of physical and mental sustenance to a large proportion of an even larger humanity—and it will probably happen sooner if we don't consider such advances as flying to be "unnatural" ("If God wanted us to fly, he'd have given us wings").

Another example of how our only feasible alternative is to keep forging ahead with new knowledge rather than to try to get closer to the "natural" state (i.e., untouched by humans) or to maintain a stable state: would you oppose developing new antibiotics that kill microbes that foil existing antibiotics? Calling a moratorium on drug development will increase the death rate. Discarding all the newfangled artificial drugs will increase it even more. Of course the new drugs bring new problems, too, such as even more resourceful microbes. But which way points to a better future a century from now?

How does population growth fit in? More people put more demands upon the system and therefore "disturb" it more. But more people also bring about greater understanding of the system and increased capacity to bend it to our will. This results both from the increased demand for goods and services that more people cause, which leads to improved technology, and from the greater potential for knowledge creation that more people represent (all else equal). The difference is between a world of 10,000 people many millennia ago, who inadvertently disturbed little and could intentionally alter little, and our many billions now who disturb much but who can purposely alter even more.

6. The Old-Time Journalistic Methods Don't Work Here

The media disseminate wrong information on population, environment, and other topics, in part, because of the scientific nature of the information. This chapter describes how the techniques that journalists use to report on traditional topics such as politics are inappropriate for scientific questions regarding natural resources and the environment.

Anecdotes versus Data

Schools of journalism currently teach nothing but the standard reportorial methods of interview and adversarial discussion of the facts. These methods serve well for covering fires and politics but fail badly on other stories modern reporters cover regarding policy issues that require the use of probability, sampling, and inferential statistics.

Consider this typical example from the *Washington Post*:

> Last month, an Atlanta newspaper columnist named Lewis Grizzard came here for a visit. He wishes he hadn't. Reason: our wondrous, knowledgeable, kind, efficient taxi drivers.
>
> Grizzard took four taxi rides during his stay. The first driver took him to Union Station when he had asked to go to National Airport. The second charged Grizzard $10 to drive around in search of a restaurant that they never found. The third ran up $11 looking for the Key Bridge Marriott—having started all of 200 yards away, in Georgetown. The fourth couldn't locate a rather well-known local landmark, the U.S. Capitol.
>
> When Grizzard got home, he wrote a column that was more like a fragmentation grenade. Conclusion: Washington is the worst taxi city in the country. . . .
>
> So I thought I would go to Atlanta for a day, take as many cabs as I could between my flight in and my flight out,

> and see whether Grizzard's home town cabs were worse than ours.
>
> It wasn't even close. [The author goes on to describe nothing but good experiences in Atlanta.]
>
> Ladies and gentlemen, the loser, and still champion: Washington, D.C.[1]

The story may seem harmless and humorous, but it was written in connection with consideration of restrictions on the number of taxi licenses in Washington by the District of Columbia government, a move that would have major consequences for the public.

Now compare this from *The American Spectator*:

> Anyone who has ever taken a cab from the Atlanta airport will be exceedingly reluctant to do so again. For years, it has been unusual to find a cab driver at Hartsfield [Atlanta] who could speak English, make change, or find his way around town. And the cabs themselves were, if anything, worse than the drivers (Norman 1988, 24).

The issue of whether regulation affects the quality of cab rides is an important policy question, but a journalist who takes a few taxicab rides in each city cannot possibly reach a valid conclusion about whether taxi service is worse in Washington, D.C., than in Atlanta. These stories are interesting, but they are dangerous. True, investigation of every question should *begin* with first-hand observation—actually looking inside the horse's mouth to check out the dental situation. But the investigation should not *end* with first-hand investigation in many cases.

To be fair, articles sometimes appear that apply solid data and analysis to these pseudo-trends, e.g., "Battered-Truth Syndrome: Hyped Stats on Wife Abuse Only Worsen the Problem" (Armin A. Brott, *Washington Post*, July 31, 1994, p. C1) and "Debunking the 'Day of Dread' for Women" (Ken Ringle, the *Washington Post*, January 31, 1993, p. A1). But this is no excuse for the worthless stuff to have appeared in the first place—

[1]Bob Levey, *Washington Post*, Feb. 16, 1987, p. C17.

and it is exceedingly rare for the rotten stuff to meet its comeuppance in such a debunking article.[2]

A valid judgment requires a representative sample of rides taken by a representative sample of customers, a sample large enough to allow for the considerable *variability* from ride to ride. Especially dangerous to truth is the practice of getting a few bad or good rides by chance and then generalizing to taxi service as a whole. Scientific discipline, and the sampling techniques that are part of it, must be brought to bear. But that discipline is not part of a reporter's armamentarium.

The representative sample need not be huge or technically fancy. A show of hands in a college class of 200 students in each city, asked whether their most recent cab rides were bad, provides an acceptable estimate of the proportion of bad rides in a city. The resulting comparison would be valid whereas the journalists' method would not.

One simple rule to avoid some atrocious blunders is that a few horror stories do not constitute evidence of a nationwide trend. This rule would save a writer from a front-page *Wall Street Journal* article asserting that "Abuse of the Elderly by Their Own Children Increases in America" (Feb. 3, 1988). Such stories are often buttressed (as was true in this case) by some statistics that seem impressive but that do not, upon examination, provide any evidence for the supposed trend. The *Journal* article does not contain a scintilla of evidence for a trend, or even an estimate of the amount of such abuse in the present. Nor would an editor of the slightest evidential sophistication let pass a piece like it even on the back page, and certainly not in column 1, page 1.

When Are Scientific Techniques Necessary?

The key to avoiding bad inferential practices is knowing when the special discipline and techniques of science are necessary. The reporter should know that scientific sampling is

[2]A wonderful article by columnist Steve Twomey in the *Washington Post* (July 6, 1995, p. B1) takes politicians to task for basing pieces of legislation on a few anecdotes heard from constituents. Yet big-time journalists make a living on stories that invariably begin with an anecdote—as did Twomey's own excellent story!

necessary to get a valid answer about comparative taxi service, and interviewing will not suffice in that case, though the interview technique is valid for covering a fire. More generally, the journalist must know the characteristics of situations that indicate the need for scientific discipline or instead indicate that the reportorial method will suffice.

The reporter's prayer should be a twist on Marcus Aurelius and Alcoholics Anonymous: Give me the reporter's skill to deal effectively with the situation in which the reportorial method is appropriate, the courage to eschew the method where it is not appropriate, and the wisdom to know which situation is which.

Scientific discipline is necessary when the chunk of the world you wish to understand presents a complex, varied, off-again-on-again picture that includes data dispersed over time or geography. Scientific methods are not necessary to describe a simple, tight, immediate, local, cause-and-effect pattern. Estimating mortality in the country as a whole requires scientific census-taking techniques, whereas finding out who died in a fire does not. Learning the effects of last week's heat wave on ice cream sales needs no special methods, whereas establishing whether there has been a rise in the earth's temperature, and whether the summer of 1997 was unusual because of the greenhouse effect, requires statistical techniques not known even to many climatologists. You can see whether there are cockroaches in your kitchen without any special equipment, but determining how many bacteria there are in the water requires a microscope.

Or consider whether brushing your teeth horizontally is more effective in reducing gum disease than brushing vertically. The important effects do not occur until months or perhaps years afterwards, and they may also vary greatly from person to person. Only a carefully controlled experiment on two samples of subjects chosen randomly from the same population can provide an adequate answer. Theoretical reasoning and even short-run observation of one or both groups are almost surely inadequate to provide a valid answer.

When immediate observation is insufficient, and when experts and libraries do not yield the needed answers, one must turn to scientifically disciplined research. And when I say "must" I mean that failure to use scientifically sound methods means that the author is fooling himself into potential difficulty, or fooling others with results that will be fraudulent at best and disastrous at worst. Please note, however, that "scientific" does not mean *experimental*. Astronomy and population censuses are two important examples of scientific enterprises that do not use experimentation. Experiments have great advantages when they are feasible, but when they are not feasible other methods are usually available. Nowadays newspapers recognize that scientific polling methods are required for useful forecasts about the outcomes of elections; gathering opinion the old way in bars and barbershops is not enough. But in too many other cases, journalists still barge ahead without the necessary scientific techniques.

Consequences of a Lack of Science

The failure to use scientifically sound methods is an increasing affliction upon the public as newspapers and television become more influential, and as we increasingly attend to events that we cannot check ourselves. The public is systematically misled about such issues as the extent of welfare abuse by immigrants, the dangers of nuclear plants and nuclear waste, and trends in the availability of natural resources and the cleanliness of our environment, because journalists apply to these issues the same techniques that work well in covering warehouse fires and trials of corrupt politicians. But these nonscientific techniques systematically provide unsound answers to the more global questions.

Was it Hegel who said that the only lesson of history is that we forget all history? One of the few negative general trends in a world where most things are getting better is our greater propensity to disregard history with each successive decade and improvement in communications, along with an increase in the volume of printed material available. For example, the average age of articles cited in scientific research nowadays is less than it was in earlier times, simply because there are more articles being written by more scientists. This means that great work of the past gets forgotten sooner in favor of trivial recent work.

Perhaps for this reason, and perhaps also because of the occupational focus on "news," journalists often devote much attention to recent changes that may be only blips running against long-run changes. For example, every few years there is a spate of news stories about how the food situation is getting worse, when in fact there is only a temporary reversal. Or a single hot summer is viewed as portending major climatic changes.

Substitutes for Science: Opinion and Combat

Journalists rely on two substitutes for scientific research: their own beliefs about what must be true and adversarial combat between quoted experts. Journalists themselves identify and explain phenomena that are beyond the powers of the human mind to achieve without the systematic procedures of science. In his autobiography, James (Scotty) Reston described how he came to the point of leaving the *New York Times* if he was not allowed to go beyond the reporting of facts. "We were, I constantly insisted, telling the reader what happened but not why" (1991, p. 121). To their credit, "The editors conceded that more explanation was necessary as the nation became more involved in world affairs, but feared, with some reason, that this might dilute and even corrupt the news columns with opinion" (p. 124).

Reston won his battle with his bosses, and he suggests that he opened the doors to others also doing what one *Washington Post* columnist, William Raspberry, calls "spotting trends" (Nov. 4, 1994, p. A25). This then spawns what *Washington Post* media columnist Richard Harwood calls "The Trend Explosion Industry" (Aug. 1, 1994, p. A21): debunking the trends that others "spot," but using the same techniques—unaided thought. Other commentators speculate that the possession by journalists of graduate degrees rather than just high school diplomas, thereby leading them to regard themselves more as advanced thinkers than simply reporters of the facts, contributes to this "trend" (which I blushingly admit may rest on factual foundations not much more solid than those I criticize).

Journalists defend their standard practices on the grounds that they enlist experts to thrash out the scientific questions. The theory is that an adversary process like that of the courts winnows out the truth. Even at its best, however, an adversary process has grave shortcomings in scientific matters. And the adversary process as practiced by journalists is not nearly as good as the process in courts.

It is ironic that journalists are often critical of the jury system's ability to deal with technical questions, but they are confident of their own ability to do so. For this to be true, a lone journalist would require more judgment than a panel of twelve jurors, because the jury has major advantages not available to journalists.

The journalist must pick his own experts. Even with the best will, he has fewer resources to do this well than do the two sides in a legal dispute. Picking experts requires expertise and considerable effort to dig into a new field. The journalist's Rolodex method is simply an accumulation of celebrity experts—a celebrity being a person well known for being well known—and often the celebrity expert expresses views out of step with, and repugnant to, the celebrity's entire profession. The egregious case of Lester Brown versus the consensus of agricultural economists is a case in point.

The journalist has less incentive to pick the best experts than do the sides in a court case who seek to find the most effective advocates for their sides. Indeed, the most effective advocates for the two sides can make a newspaper story seem inconclusive, which no journalist wants. *Newsweek* headlines "More Bad News for the Planet" (March 28, 1988, p. 63). An expert who maintains that on balance there is more good news for the planet than bad news is hardly welcome at the funeral feast. And since there is no check on the journalist's picking experts in such fashion, the overall impression given by an article is the one the journalist favors.

In court, rules apply equally to the two sides' experts, giving equal opportunity to both. But a journalist makes the rules and applies them at the same time, a process that journalists themselves would criticize in any other venue. And the journalist elicits the experts' views with one-on-one interviews, without the bracing effect of cross-examination by other experts.

Perhaps most infuriating to the experts themselves, journalists often consider themselves qualified to render summary judgments in matters where there is controversy rather than consensus among the experts. The serene arrogance of such practices will pass without further comment.

The heart of the jury system is that the individual jurors have no personal stake in the outcome. But journalists do have a built-in bias—a bias for the slant that is more newsworthy. This bias must affect the choice of experts, and the choice of what to quote from the experts. So for this reason, too, the combat method for ascertaining truth does not protect journalists from the hazards of writing stories about events that require scientific discipline without exerting that discipline themselves.

What Will Change the System?

What forces will make for improvement in the quality of the "news" we read about on topics that require scientific discipline to understand them well? The explanation of poor performance is the incentive structure. It therefore is foolish to blame individuals. One must try to think of ways to improve the structure. To point this out is not press-bashing; it is the ordinary economics of management.

I cannot forecast the causes or course of such improvement; only time will tell. The federal Food and Drugs Act of 1906 saved the public from some snake-oil excesses, and later legislation improved the situation even more, though the movement may now have gone too far. But we cannot expect or hope that legislation will improve the truth level in journalism about social-scientific topics; the solution might be worse than the problem even if it could ever be adopted.

In the absence of legislation, progress will be slow, because nothing on the horizon promises to help much. Yes, editors more knowledgeable about social science might help. In-house scientific ombudsmen, or a board of outside social-science ombudsmen serving for short periods, perhaps in rotation, might also help. It is doubtful that the courts can help; who would sue whom on alleging what harms done? But one or another improvement must eventually come, as it came with adulterated foods and false drugs.

7. Damn Lies, Statistics, and Doomsday

This chapter arrays some of the statistical contrivances that are used to paint false pictures of various social issues. As case studies I'll use a widely distributed 1993 pamphlet entitled *The Index of Leading Cultural Indicators* by a widely known public figure in the United States, William Bennett, author of the best-selling *The Book of Virtues;* Vice President Albert Gore Jr.'s book entitled *Earth in the Balance;* and a *Washington Post* front-page story about blacks in the United States. And though those particular publications will be quite forgotten in a very few years, the material in these case studies will remain evergreen because other publications in the future will use the same contrivances for the same dishonest purposes.

The Booklet of Statistical Vices

My purpose is not to attack William Bennett personally. Rather, I want to use this as an educational case study of the abuse of statistics. To help in the analysis I'll use as a cookbook a text entitled *Statistics: A New Approach* by W. Allen Wallis and Harry Roberts, written in 1956 and a landmark in statistics education for decades. One of its central features is instruction in how to avoid pitfalls in statistical thinking, and it lists many of the most common fallacies.

Bennett says in his introduction (p. 1) that "we have experienced substantial social regression . . . over the last 30 years." He goes on to assert that "there has been a 560 percent increase in violent crime; more than a 400 percent increase in illegitimate births; a quadrupling in divorce rates," and so on. Let's see the quality of the evidence he provides for his argument.

"Number of Crimes Committed" is the first topic Bennett addresses (p. 2). He presents data about "Millions of Violent Crimes" and "Millions of Total Crimes," both since 1960. Here

Bennett commits the first in Wallis and Roberts' list of "Misuses of Statistics." He makes no adjustment for the growth of the country's population since 1960. Therefore crimes *per capita*—the relevant measure—have not grown as fast as Bennett's charts show.[1]

The next abuse is the choice of the time period. Why since 1960? If Bennett had plotted a much longer period backwards, we would see that crime had been vastly higher at some times than in 1960. The date he chose coincides with his conclusion better than most other dates would.

The really crucial issue for the purpose at hand is the source of the data. Bennett's numbers come from the FBI. But informed social scientists know that FBI data refer only to *reported* crimes. The rate of reporting has increased greatly in recent decades, resulting in an increase in reported crime.

Since 1973 the U.S. Bureau of Justice Statistics has conducted "victimization surveys." These surveys show no growth in violent crime over the period of the surveys and a large decline in nonviolent crimes such as larceny, burglary, theft, and auto theft. But Bennett presents only the scary FBI series that seems to support his general claim even though the victimization data are available in the *Statistical Abstract of the United States*.

Bennett's second topic is "Median Prison Sentence for All Serious Crimes." His graph shows a decline from 25 days in 1954 to 5 days in 1974, and then a modest rise to 1990, suggesting that the society has gotten soft on punishing crime. But can *five* days in prison really be the "median sentence" for serious crimes? Of course not. So what do the numbers in Bennett's graph mean? The numbers do not refer to the "median sentence" that is in the title, but rather are a computation called "expected punishment" for the commission of a crime, a computation that includes the likelihood of being arrested, the chance of going to trial, the chance of being convicted, the chance of getting any prison sentence, and the sentence itself. Expected punishment is a very interesting and

[1]Bennett notes this in his discussion, but that's like the small print in a contract. It's the dramatically rising curves in the graphs that hit the reader over the head.

useful concept, but it has almost no connection with what Bennett claims to be talking about. This is an abuse of statistics akin to the nonsequitur fallacy in rhetoric.

Another common statistical misuse is combining apples and oranges when they don't belong together—the "fallacy of composition." Does "serious crimes" in the "median sentence" graph have the same composition (of crimes)—and hence the same meaning—over the decades? Certainly not. Homicide, rape, and theft have changed at vastly different rates, so lumping them together produces confusion at best.

Why do the prison sentence data start in 1954 while the crime data start in 1960? The obvious answer is that starting this graph at 1960 would make a less dramatic chart. The presentations are rigged to achieve the worrisome effect that the author wants to achieve.

Bennett's third topic is "Juvenile Violent Crime Arrest Rates." Here, at least, Bennett presents the data on a "per 100,000" basis. But he does not tell us what the "100,000" refers to. Juveniles? Population? One cannot know what mischief may lurk behind the undefined number. Vagueness of definition is one of the most useful practices for the statistical double talker.

The juvenile discussion contains all the fallacies described earlier with regard to total violent crimes and some additional fallacies. The vertical axis of the graph used to display the data does not start at zero. This is a real meat ax of a crude statistical trick, grade school stuff. If the vertical axis had run all the way to zero, the rise in the curve would have been less dramatic to the eye.

The subject is *arrests* rather than *crimes,* though the casual reader is not likely to notice the shift. Is it possible that juveniles have been getting arrested more frequently for given crimes than in the past? We don't know. But if so, the rate of arrests would give a misleading impression of the amount of crime. Shifting the definition is a most useful contrivance for portraying a false statistical picture.

The fourth topic in Bennett's pamphlet is "Children Relying on AFDC" (Aid to Families with Dependent Children). Again Bennett uses the now-familiar trick of showing total numbers of children without adjusting for population increase.

Bennett also mistitles the graph for AFDC as "Relying on," when the data refer to those who *receive* AFDC. Who knows how many children get AFDC who do not rely on it at all? Indeed, how many children received AFDC half a century ago? How many children receive AFDC-type payments in Somalia? Zero in both cases, even though the children were and are more needy in those cases. The explanation, of course, is that the figure may well show an increase in the generosity of the generous government programs rather than greater need or reliance.[2]

The rest of the booklet is depressingly similar. It also includes a howler. The rate of births per thousand teenagers is drawn in Bennett's graph as if it has risen from 15.3 in 1960 to 42.5 in 1990. But the official data from the National Center for Health Statistics show that the birth rate for teenagers (aged 14–19) fell from 89.1 in 1960 to 59.9 in 1990 (including an upturn since 1986). Bennett must have made some extraordinary error.

No wonder the public distrusts statistical presentations. It is easier to mislead with numerical presentations than with words alone because there are more subtle tricks you can play. The only good that comes out of Bennett's booklet is that I have a wonderful demonstration of the statistical abuser's art for regular use in my elementary statistics course.

William Bennett probably would not be called a darling of the press. And journalists pride themselves on their "hard-nosed" skepticism. But I have not read any criticism of Bennett's statistics in this booklet. And this is not because the subject would take tedious, difficult digging. I wrote the pages in this chapter in about three hours, with the assistance only of the *Statistical Abstract of the United States*, which should be on every journalist's desk. How come Bennett's falsities got a free ride? Is it because journalists simply share Bennett's general belief about "social regression" and therefore don't even subject the data to scrutiny?

[2]Indeed, this is suggested by the rapid rise in AFDC receipt shown between 1960 and 1975.

Truth in the Balance

Albert Gore Jr.'s book is called *Earth in the Balance*. But it is truth that is in the balance, rather than our very durable planet. The book is as ignorant a collection of clichés as anything ever published on the subject. And there is much tough competition for that abysmal bottom spot.

Just about every assertion in the book points in the wrong direction—suggesting that conditions are getting worse rather than getting better, which they are. Lest the reader accuse me of hunting-and-picking for errant soft targets, let's start with the very first topic in the book, soil erosion, and go from there.

After the obligatory drama about how "eight acres' worth of prime topsoil floats past Memphis every hour," Gore says that Iowa "used to have an average of sixteen inches of the best topsoil in the world. Now it is down to eight inches."

The first footnote in the book says only that his source was "conversations with the U.S. Army Corps of Engineers and the Iowa Department of Agriculture and Land Stewardship." One cannot check the Iowa situation with this anonymous quote. (Indeed, printed sources are generally scarce in the book.) But we do know the trend of increasing erosion for the country as a whole.

If Gore had done his homework, he would have examined the data in the publications of the U.S. Department of Agriculture's Economic Research Service. He would have talked to Bruce Gardner, now assistant secretary of agriculture for economics, and to Gardner's teacher at the University of Chicago, Theodore Schultz, who has been watching soil erosion since his days as a farmboy in South Dakota in the 1920s. Schultz received a Nobel prize for his work in agricultural economics and human capital. Gore would have read the articles by Schultz and by respected agricultural economists Earl Swanson at the University of Illinois and Earl Heady at the University of Iowa.

Gore would then have found that the facts are exactly the opposite of what he writes. The farms in the United States are becoming less rather than more eroded, on average. Decade after recent decade, fewer rather than more acres suffer from severe erosion. This emerges from comparison of Soil Conservation Service surveys done at intervals since the 1930s.

The second specific item Gore mentions is DDT, "which became for me a symbol of how carelessly our civilization could do harm to the world." He gives no data and provides no references, though he later adds that DDT "can be environmentally dangerous in tiny amounts." A touch of research would have turned up tons of writings such as *Mosquitoes, Malaria, and Man*, by Gordon Harrison, who was director of the Ford Foundation's environmental program. Gore would have learned that with the aid of DDT, "India had brought the number of malaria cases down from the estimated 75 million in 1951 to about 50,000 in 1961. Sri Lanka . . . reduced malaria from about three million cases after World War II to just 29 in 1964." Then as the use of DDT went down, "Endemic malaria returned to India like the turnaround of a tide." By 1977 "the number of cases reached at least 30 million and perhaps 50 million." Does that suggest that DDT does harm to civilization or does good?

Gradually, it became clear, too, that DDT could be used quite safely. The scary scenarios in Rachel Carson's book, *Silent Spring*, which Gore remembers troubled his mother, turned out to be without foundation. In 1971, amidst the fight that led to the banning of DDT in 1972, the president of the National Academy of Sciences—distinguished biologist Philip Handler—said, "DDT is the greatest chemical that has ever been discovered." Commission after commission, top expert after top Nobel prize-winning expert, has given DDT a clean bill of health, as Gore could find out in Elizabeth Whelan's *Toxic Terror* (1993) and the host of references therein. But evidence on such matters has no place in Gore's book.

The third item Gore mentions is Agent Orange, which he uses weasel words to describe as "the suspected cause of chromosomal damage and birth defects." Again, no references in Gore's book. And with reason, because though Agent Orange (dioxin) was indeed "suspected" by Gore and his colleagues, it was pronounced innocent in the federal courtroom when veterans brought suit. There simply is no solid scientific evidence of ill effects from dioxin.

In August 1991, the *New York Times*'s front-page headline was "U.S. Backing away from Saying Dioxin Is a Deadly Peril." The story continued, "Exposure to the chemical, once thought to be much more hazardous than chain smoking, is now considered by some experts to be no more risky than spending a week sunbathing." And the Centers for Disease Control now admits that the Times Beach evacuation was unnecessary. But Albert Gore has not gotten the word.

Love Canal is next in the book. Gore seems unaware that the solid scientific consensus is that there was no observable damage to humans from living near Love Canal. So far we have only reached page 3. And the entire book is filled with this sort of environmental gossip, backed by no sources, and contradicted by solid data.

Though Gore undoubtedly cares sincerely about environmental and resource issues, his ignorance is willful rather than naive. He has been told in the past that his utterances on these subjects do not correspond with the facts. But he has chosen to ignore the scientific literature.

Furthermore, the advisers Gore leans heavily on—Paul Ehrlich and Lester Brown—have been proven wrong in every one of the forecasts they have made in the past two decades, a truly astonishing record of consistency. Yet it is still their agenda that Gore puts forth, almost as if he is writing from handouts of the environmental movement.

Moreover, Gore is suspicious about others' motives and behavior. He writes,

> The statistics about forests can be deceptive too: although the United States, like several other developed nations, actually has more forested land now than it did a hundred years ago, many of the huge tracts . . . have been converted from diverse hardwoods to a monoculture of softwood.

But the same U.S. Forest Service statistics that showed Gore that the total volume of trees is increasing also show that the volume of hardwood trees is going up, rather than being driven out by softwoods. Just who is deceiving whom?

The reader may wonder who is to be believed. One crude test is whether people will put their money where their mouths are. So here is my offer: I'll bet a week's or a month's pay with Mr. Gore or anyone else that I've got the above matters right and he does not. And I'll go further: I'll bet that just about any broad aggregate trend pertaining to human welfare will improve rather than get worse—health, standard of living, cleanliness of our air and water, natural resource availability—you name it, and you pick any year in the future. First come, first served.

It is not surprising that a senator (as Gore was then) does not have time for the kind of library digging that an academic researcher does. But that is no excuse for publicizing and acting on wrong facts, because it is not harmless. Gore would (among other measures) tax the use of new raw materials to force more recycling, establish higher mileage requirements for cars, require "efficiency standards throughout the economy"—all of which would raise costs and increase government intervention in people's lives. All this on the basis of beliefs he holds that are utterly contradicted by the solid scientific facts.

Racial Worry and Frustration Built by the *Washington Post*

The headline across the first and second columns of the front page of the *Post* on October 10, 1995, was "Worry, Frustration Build for Many in Black Middle Class" (Merida 1995) and continued in the second and third paragraphs: "Many middle-class blacks are more jittery about their futures than they've ever been. . . . The escalating anxiety among the black middle class was one of the findings of a national survey sponsored jointly by the *Washington Post*, the Kaiser Family Foundation, and Harvard University."

The article reported the poll results, and even gave the "methodology." But I found no backing in the poll data for the statements quoted above. There was nothing that showed trends over time or that there had been a change in any direction.

On the only questions that refer to the past, the responses for whites and blacks are almost identical; in fact, blacks are slightly *less* convinced that things have gotten worse. The scores in the block of questions entitled "I believe, compared to 10 years ago" and then "I am farther away from attaining the American Dream" were: whites 62 percent, blacks 60 percent; "That it has become harder to get good jobs": whites 55 percent, blacks 53 percent; and "That it has become harder to find affordable housing": whites 57 percent, blacks 52 percent. Hence whatever was true of blacks was true of whites, on this evidence. But the point of the story is that the experience of blacks is different—and worse—than whites.

Furthermore, there is a large body of scientific research showing that questions comparing the past to the present are almost always invalidated by the "good old days" bias. Hence the reported data cannot reasonably be interpreted as any evidence at all about changes over the years in the thoughts of blacks or whites.

I called writer Kevin Merida and asked whether he could supply further data, but he said he could not. He said that the interpretations of trends were based on a variety of perceptions by various people. He also noted that as the writer he was not responsible for the headline.

I also called Richard Morin, who is responsible for *Post* polling. He referred me to Merida and said he believed Merida was relying on data from a recent book by Jennifer Hochschild (1995, p. 57, Table 3-2, pp. 62–63). I said that I had spoken to Merida and Merida had not mentioned these sources. I also asked Morin to send me xeroxes of relevant Hochschild data—a reasonable request among social scientists—but he suggested I consult the book.

I went to the trouble of examining the evidence in Hochschild's book, *Facing Up to the American Dream*. But there I found no support whatsoever for an increasing trend in "jitteriness." Hochschild says flatly on page 57, "Members of both races are as sanguine now as they were four decades ago."

Furthermore, she shows at length in her Tables 3-1 and 3-2—referring to exactly the same sorts of questions in the *Post* poll, but asked in studies from the 1960s to the 1990s—that blacks are invariably more optimistic than whites. Indeed, the *Post*'s own poll found just that, as cited above. All this is exactly the opposite of what the *Post* story suggested.

I faxed this material from Hochschild's book to Mr. Morin on October 15, 1995, but I have not heard from him.

If politicians or commercial enterprises made assertions such as the *Post* made in this story, either without supporting evidence or in actual contradiction to the evidence, the *Post* would pillory them or call for their prosecution. But no law protects the public from a newspaper's doing this sort of thing, and indeed, the press always protests that any such legal action would "chill" a free press.

Indeed, the press has a special responsibility for accuracy in these matters because it righteously takes objectivity and truth as its guiding standards, which neither politicians nor businesspeople claim for themselves. Yet the *Post* feels free to proceed as it does in this and in dozens of other similar cases of false bad news that I have collected and sent to editor Leonard Downie and to the ombudsman, the supposed protector of the public in such affairs, but with no response. Nor will another newspaper enter the fray and criticize its competitor in these respects; this code of silence closely resembles the behavior of the Los Angeles cops, so criticized by the *Post* in connection with the O. J. Simpson trial.

As to nonjournalistic writers, who among those who would like to have his work printed from time to time in the *Post* is suicidal or stupid enough to want to address the *Post* on such matters as these? And no advocacy organization has a stake in raising a howl about such misstatements; no one's interests are harmed except those who care about the plain, simple truth.

8. Personal Knowledge versus Media-Shaped Opinions

Polls that ask individuals about both their *own* situations *and* the rest of the nation or world at large provide evidence that people receive from the press and television an overnegative impression of the world around them. A striking and consistent discrepancy exists between the poll results of people's assessments of what they know personally and what they think is happening "out there."

People have much more favorable views of what they *know first hand* in their own lives than what they think is *the situation elsewhere*. As Lipset and Schneider put it, "Americans repeatedly express optimism and confidence about their own lives and their personal futures, even while decrying the terrible mess the country is in" (1987, p. 8). And since the country is (from one point of view) the composite of its citizens' lives, these data show a systematic negative bias in people's assessments of the situation at large. That is, if there were no bias and people were to judge accurately about other people's situations, the average of people's judgments about their own situations would equal the average judgment about the country as a whole. But the two are *not* equal, showing a downward bias in the abstract "out there" assessment.

The main source of information about most other people's situations and lives is the media. Therefore, it is reasonable to assert that the media are responsible for this negative bias in people's assessments about the situation of the country at large. Lipset and Schneider comment on this phenomenon as follows: "A striking characteristic of the decline of confidence is that it is almost entirely related to events beyond people's own personal experience: conflicts, scandals, protests, and failures that affect their own lives indirectly, if at all" (p. 8).

But why should there be a decline with respect to public events but not to private events? Perhaps the explanation is that the press has come to pay more attention over the years to "conflicts, scandals" and the like. And perhaps this is because with every passing year there is less of other news—fewer wars in which the United States is engaged; fewer catastrophic natural disasters because of better predictions; fewer disease epidemics; fewer catastrophic fires; and fewer other events that threaten life and limb and which have been major worries throughout human history.[1]

Other evidence comes from the comparison of the present to a few years earlier. Individuals see the present as better than the past in their own lives, which is an accurate assessment on average because health and the standard of living have improved over time; people can know this from first-hand evidence, and they imagine that the future will be better for them than the present. But with respect to the country as a whole, and to the economy, about which they cannot know first hand, they regard the past as better than the present—the usual "good old days" nostalgia. This discrepancy was shown typically in a 1991 poll about whether people felt "Not as well off" as three years earlier. "You yourself are" not as well off received 33 percent of responses, but "Most Americans" received 48 percent of the responses, compared to "Better off" and "In about the same shape."[2]

The same effect appears when the subject is the environment. When asked about the environmental conditions *in their own area*—whose conditions they know personally—as well as conditions in the country as a whole, respondents rate the local environment more highly than the environment in the country as a whole, and indicate a much lower degree of worry about it (Simon 1981/1996, Figure 5-2). When asked before Earth Day 1990 whether pollution is "a serious problem

[1]The best measure of this is the rising level of life expectancy in the United States and in the rest of the world.

[2]ABC News/*Washington Post*, October 18–21 and December 11–15, 1991, in *The American Enterprise*, January–February 1992, p. 99.

that's getting worse" for "the country as a whole," 84 percent said "serious," but with respect to "the area where you live," only 42 percent said "serious."[3] As the *Compendium of American Public Opinion* put it, "Americans are primarily concerned about the environment in the abstract . . . most Americans are not worried about environmental problems where they live . . . most Americans do not feel personally affected by environmental problems."[4] The average person feels that the grass is greener on *his own side of the street*—or more precisely, that the grass is browner on the other person's side of the street, which the comparer has never even seen. Again, this cuts the logical ground out from under the abstract aggregate judgments, because they are not consistent with the sum of the individual judgments.

The same phenomenon appears with respect to immigration. Even though over the decades all polls have shown that Americans do not favor more immigration when asked about it in the abstract, Americans have positive feelings toward the immigrants *in their own areas*, and toward the immigrants they know personally. The comparison between the results of the two inquiries can be seen clearly in a 1978 poll about Vietnamese immigrants. When asked, "Thinking now about the Indochinese refugees, the so-called 'boat people'; would you favor or oppose the United States relaxing its immigration policies so that many of these people could come to live in the United States?"—32 percent of the respondents were in favor, 57 percent were opposed, and 11 percent had no opinion. But when asked, "Would you, yourself, like to see some of these people come to live in this community or not?"—48 percent said "yes," 40 percent said "no," and 13 percent expressed no opinion. There is an interesting split in thought here, with the greater voiced opposition apparently being based upon general belief formed by the mass media, and the greater voiced support coming from personal experience with immigrants (R. Simon 1985, 42).

[3]CBS News Poll, April 16, 1990.

[4]Gilbert (1988), pp. 121–22.

People are more likely to have a rosy view of "the good old days" with respect to the overall society, with which they do not have first-hand experience, than of earlier times in their own lives. They have personally experienced the secular improvements in health, standard of living, and the like, and hence they accurately report improvement. The fact that they have a different view of that which they do not know, and a less positive view of the present compared to the past, can be attributed only to their sources of second-hand information (or supposed information).

How Can It Be That People Think What They Do?

My mother was born in 1900. One of her brothers died of diphtheria in infancy, while the doctor looked on helplessly. In 1937, her only son was saved from death at age 5 by the first new wonder drug, sulfanilamide. In her 80s Mother knew that her friends had mostly lived extraordinarily long lives, usually in good health. She was grateful for the new miracles of medical science, and she appreciated the convenience and comfort provided by such modern inventions as the telephone, air conditioning, and airplanes.

Yet Mother insisted that life was worse in the 1980s than it was when she was young. When I pressed her why she thought so, she said, "The headlines in the newspaper are all bad."

Similarly, when I occasionally say to my wonderful Aunt Ruth, now in her 80s, that pollution is decreasing in the United States, she responds, "But the pollution in the bay [near her home in Queens, New York] is much worse than when we moved here." When I remind her of the pollution in the drinking water from the nearby Hudson River that killed children with typhoid and diphtheria when she was a girl, she sighs, "I guess you're right," but I don't think I have changed her outlook much.

When I say to my wonderful Aunt Anna, also in her 80s, that everything material is better now than when she was young, she answers, "But you read in the papers about so much wrong-doing." Reminding her of the horrors of two world wars—when paradoxically good news often was featured in the papers—and mentioning the frauds of the 1920s, brings her to nod agreement, but that's only because she loves her nephew, I think.

Lest one think that those who are more involved in the business of the world than my mother and aunts are, and whose opinions shape events, somehow have the ability to discount false bad reports and are able in some mysterious prescient manner to peer into the heart of truth, consider this incident reported by journalist David Broder. Broder certainly is as much in touch with current events as any human being on the face of the earth, and is much respected for his supposed clear understanding and fair and balanced reporting. Following a trip to Europe in the spring of 1993, Broder wrote this about trying to understand the attitudes toward the United States of the attendees at a conference of movers and shakers from various countries in Europe and Asia:

> You have to make a mental adjustment that I found difficult. You have to see the United States, not as most Americans do, as a nation beset by problems and maybe headed down the chute, but as a citadel of economic and political strength in a world of stumbling economies and faltering leaders.
>
> It is startling to be told that no major economy is growing as fast or generating jobs as well as the United States is today. But the figures are irrefutable.[5]

If David Broder can entirely misunderstand whether conditions had been getting better or worse in the United States and can so misjudge the overall standing of the United States relative to other countries, should we be surprised that less-informed persons also have things backwards and upside down?

More General Evidence

The power of the media to produce negative judgments concerning abstractions of which the public cannot possibly have an informed judgment is shown by polls about "the most

[5]*Washington Post*, June 9, 1993, p. A19.

important problem facing the country today." In 1987, 1988, 1989, and 1990, the "budget deficit" received either the highest or the second-highest percentage of votes.[6] The national budget balance is an extraordinarily complex issue about which there is much controversy even among economists (who tended to be less concerned about this matter than noneconomists, it seemed). Yet strong public judgment was elicited by the media reports.

In a 1996 poll, 70 percent of the respondents said that the federal budget was "larger" as "compared to five years ago" when in fact it was less than half in absolute terms and much less than half in percentage terms.[7] The front-page headline was "A Nation That Poor-Mouths Its Good Times." Much the same is true of the aggregate unemployment rate (estimated by the public at four times its actual rate) and the rate of inflation. It is most welcome that there is mention of the fact that the times are good, even if the emphasis is on "poor-mouthing" the facts. But the news story never even hints that the only basis for any attitudes about the deficit is news stories; no one ever meets a federal deficit first hand.

Many years ago sociologist Richard LaPiere showed the enormous inconsistency between what people say in the abstract and their own behavior. For two years starting in 1930, he and a young Chinese couple crisscrossed the United States together, and among 251 events of eating and staying in motels, they encountered racial discrimination only once. But when those very same establishments, plus another comparable sample, were asked in a mail survey, "Will you accept members of the Chinese race as guests in your establishment?" the vast majority said that they would not (Plous 1993, 59).

Conclusion

In surveys, there is a consistent discrepancy between the public's beliefs about the environments that they know first

[6]CBS News/*New York Times* and Gallup polls in *The American Enterprise*, January–February 1992, p. 101.

[7]*Washington Post*/Kaiser Family Foundation/Harvard University poll, *Washington Post*, October 13, 1996, pp. A1, 38.

hand, and those they only know second hand. Poll respondents view the situation they know at first hand more positively than the situation at large. The only likely explanation is that newspapers and television—the main source of notions about matters which people do not experience directly—are systematically misleading the public, even if unintentionally.

There is a vicious circle here:

- The media carry stories about environmental scares.
- People therefore become frightened.
- Polls then show people to be worried.
- The worry then is cited as support for policies to initiate actions about the supposed scares.
- These policies raise the level of public concern further.

The media proudly say, "We do not create the 'news.' We are merely messengers who deliver it."[8] These data show that the opposite is true, at least in this case.

[8]Richard Harwood, ombudsman, *Washington Post*, May 31, 1992, p. C6.

9. How Psychology Affects the Evaluation of Trends

The evidence is overwhelming that the material conditions of life are getting better rather than worse. Why, then, do people have a negative perception about trends in environment and resources despite the positive evidence?

An important source of the belief that things are going poorly, and that the future outlook is gloomy, surely is the reference point for the comparisons that a person makes. Wise people have written for thousands of years that whether a person feels happy or sad depends in considerable part on the benchmark against which he compares beliefs about the current state of affairs. So important is this element in our mood that it is the key element in psychological depression.[1] The choice of benchmark for comparison is seldom forced on us by the world; rather, it is mainly within our control. The benchmarks that we choose to use in comparisons suffer from three distortions: recency bias, rising aspiration levels, and misperception.

We take for granted an improvement in our lives. We compare other states of affairs to the new and improved state rather than the state of affairs in the more distant past. And if our prospects are not better than the newly improved state, we grouse. This psychological mechanism of rising expectations explains much about people's thinking.

The development of the personal computer and people's feelings about it provides an illustration. People commonly are delighted with their first computer because they immediately see how much it eases their work. But we quickly take

[1]For a full discussion of the subject of comparisons, especially in connection with psychological depression, see Simon 1993.

the computer for granted and eventually become dissatisfied that it does not work even faster. A delay of a few seconds becomes irksome, even as the computer is saving hours or days compared to the precomputer situation. The old programs feel "clunky." The result is a continual desire for faster and faster computers, bigger and bigger hard drives, more work-saving utilities, and ever-fancier refinements on the programs we use. Some people wind up no happier in their work than before they had computers.

Childbirth provides another example. An obstetrician in Washington, D.C., announced in 1990 that he was quitting the practice of obstetrics because (for the second time) a prospective mother had asked of him the impossible—that he guarantee her a perfect baby. In years past, when life in general, and childbearing in particular, were so much riskier than now, no woman would have even thought of such a request. But now women expect and demand complete safety during labor and delivery, and flawless offspring.

A third example is provided by environmental pollution. A judgment that our air and water now are "dirty" and "polluted" is not reasonable when compared with the terrible pollutions that were banished in the past century or so—the typhoid fever that polluted even the Hudson River at New York; the smallpox that humanity has finally pursued to the ends of the earth and apparently eradicated; the dysentery and cholera that distressed and killed people all over the world, as it still does in Asia and Africa; the plagues and other epidemics that harm us much less than in generations past, if at all. Not only are we in the rich countries free of malaria (largely because of our intensive occupation of the land; see Simon 1981/1996, 460–63), but even the mosquitoes that only cause itches with their bites are now so few in many affluent urban areas that people no longer need window screens for their homes and can plan garden parties at dusk.

No matter how good things become, our aspiration level rises so that our anxiety level declines hardly at all. We focus on ever-smaller actual dangers. Parents manage to worry about their kids' health and safety even though the mortality of children is spectacularly less than in prior decades and centuries. And orthodox Jews and Muslims in the United States continue to worry about whether their food is ritually pure even though the protections against ritual contamination are remarkably better than in the past.

Once upon a time orthodox Jews said, "A Jew eats a small pig every year without knowing it." Nowadays, with plastic wrapping at the manufacturer, and the microscopic examination techniques of modern science, the level of food purity is much higher than in the past. But the level of concern does not seem to abate.

Misperceptions about the Past

One of the problems in understanding modern life is the implicit comparisons with a past that never existed. Some of us imagine Africa only decades ago as people swinging from the trees like Tarzan, pre-Columbus North America as Native Americans sitting around campfires and growing up to tall strong adulthood on plentiful, organically grown, pesticide-free food, never suffering the difficulties of adolescence; the virgin U.S. Middle West as a fertile area where people needed to do no more than throw seeds upon the ground for there to be bountiful harvests; Europeans in the Middle Ages spending most of their time dancing around the maypole and only rarely having to pull a forelock in respect to some authority; almost no one dying before old age, with death being sudden and painless; and microorganism diseases such as tuberculosis and plague being mere romantic interludes suffered by a few artists.

But these faulty comparisons are not inevitable. You can train yourself to reflect on the comparison between what you have now and what you had in the past, rather than on between what you have and what you might have, or what others have. I get enormous pleasure from having an alarm clock with a snooze bar so that I can be awakened, hit the bar, and grab another nine minutes of lovely sleep. I keep in mind how it was to work without a computer and a copy machine. And I remember how Oliver Wendell Holmes carried the manuscript of his first book with him in a satchel to every social engagement, for fear of losing it in a fire while he was away. Nor was this an imagined fear; at least one author lost forever the only copy of his novel on the New York subway. Copy machines and electronic copy relieve authors today of this nightmare.

Comparisons across Groups Rather than Time

Many discussions about public policy compare the present state of *one group* to the present state of *other groups*, as a measure of "equity." Others compare the current actual situation to the *best* possible, or to ideal purity, ostensibly to motivate improvement. A typical front-page story from the *Washington Post* (July 5, 1991) does both; it headlines a complaint of blacks that a nearby county "Isn't Drawing Upscale Stores," and the caption under a picture says "Prince George's resident Howard Stone is angered by the shortage of upscale retail stores in his community." (Yes, this article was on the front page.) This issue is very different from the sorts of problems that most of humanity has faced throughout most of its history.

A more serious example is provided by infant mortality. Consider how trends in black infant mortality are evaluated. I've asked this question of many audiences, both laypeople and professionals—even demographers. Almost everyone's reaction in the United States is that black infant mortality is a bad situation. In 1915 white infant mortality was almost 100 deaths per 1,000 births, and black infant mortality was fully 180 deaths per 1,000 births. Both rates are horrifying. And the rates were even worse in earlier years in some places—up to 300 or 400 deaths per 1,000 births. White infant mortality is currently about 9 per 1,000, and black infant mortality is about 18 per 1,000. Of course it is bad that mortality is higher for blacks than for whites. But should we not be mainly impressed by the tremendous improvement for *both races*—rates falling to about 10 percent of what they were—with the black rate coming ever closer to the white rate? Is not this extraordinary improvement for the entire population the most important story—and a most happy story? Yet the press gives us the impression that we should be mainly distressed about the state of black infant mortality.

Is this assessment balanced? What about the supposed ills and evils of today that were not foreseen in earlier times—the supposed breakup of the family; increasing numbers of homeless people on city streets; automobile and airplane deaths; children's fear of nuclear war; and other contemporary pathology (actual or mythical)? Here we must come to grips with the central premise of economic thought—that what the individual chooses is called "better." Would people from the past prefer to have had the troubles that they wished to end, and which we have ended, or the supposed new troubles of today? My guess is that when faced with the choice of wild beasts and the plague or life in the 20th century our ancestors would choose the present.

Conclusion

The evidence is overwhelming that the overall trend is toward all the material conditions of life getting better rather than worse. Yet people have a negative perception about conditions in environment and resources despite the positive trends. An important cause of the belief that things are going poorly, and that the future outlook is gloomy, surely is the type of comparisons that a person makes.

10. Why Do We Hear Prophecies of Doom from Every Side?

Another internal mechanism that affects what people believe about the past and future is the human predisposition to be attracted by prophecies of doom—and in the case of some, the propensity to make such prophecies. This chapter discusses these predispositions historically.

David Koresh and his Branch Davidians engaged in pitched gun battles with the U.S. government in 1993, leaving more than 80 men, women, and children dead as a result of apocalyptic religion. The parent Davidians religious group prophesies an end of the world in the foreseeable future. Indeed, the Seventh Day Adventist church, from which the Davidians split in 1934, was founded in 1863 on end-of-the-world prophecy.[1]

Such views are not new. The book of Revelation in the New Testament vividly warns of impending doom. And many of the Hebrew prophets forecast apocalypse.

But it is not just religious visionaries who forecast doom. Scientific discoveries are sometimes interpreted, even by nonreligious institutions, as harbingers of doom. Only five months before the Branch Davidians disaster, *Newsweek*'s cover headlined "Doomsday Science: New Theories About Comets, Asteroids and How the World Might End" (November 23, 1992). Every scientific discovery gets interpreted by some people as a harbinger of doom. "Chaos Theory Seeps into Ecology Debate . . . Disrupts Humanistic Notions of Order and Progress" was the *Wall Street Journal's* headline of a page 1 story (July 11, 1994, pp. A1, A8). In the same article Irving Kristol says, "We are at a unique moment in Western culture, the

[1]*Washington Post,* March 21, 1993, p. A18.

collapse of secular, rationalist humanism" in which "progress is the premise." An interesting, purely mathematical discovery has turned into grounds for fear of natural and social catastrophes.

The apocalyptic prophetic impulse sometimes leads to serious violence. Theodore J. Kaczynski, the Unabomber, who killed and maimed with mail bombs in the United States for many years, murdered in the name of ecological causes, stating his arguments in an anti-technology screed.[2] His case startled the public because of Kaczynski's earlier academic brilliance as a mathematician. But learning and mental agility have never prevented this kind of thinking, especially in connection with technology. Books by professors with titles like *The Surrender of Culture to Technology* (Postman 1992) have been common fare through the decades. Almost every event is seen by someone as an actual or potential cause of moral and cultural decay, as in the story about Palestinian leaders headlined "Leaders Warn of Moral Decline, But Lifted Curfew Elates Gaza."[3] Though their prophecies are less dramatic, not a week goes by that an environmental doomsayer does not warn us that "our existing world is crumbling" (Rifkin 1989, front cover). The environmental apocalypses are more gradual than the religious versions. But environmental breakdown is more widely believed—by more than half of the U.S. public, the polls tell us (see chapter 1), and by many of the leaders in every walk of life. The prophecies of environmental doom therefore deserve our attention.

Why do prophets utter their bleak prophecies? Why do people believe and act upon bleak prophecies?

Analysis of Apocalyptic Thinking

It is striking how similar sorts of forecasts of doom, together with assertions that previous times were the "good old days," have been heard in all ages. This was said to have been found on an Assyrian tablet, and it is good fun even if it is just a hoax:

[2] *Washington Post*, April 9, 1996, pp. A1, A7.

[3] *Washington Post*, June 5, 1994, p. A30.

> Our earth is degenerate in these latter days; bribery and corruption are common; children no longer obey their parents; every man wants to write a book, and the end of the world is evidently approaching (Homer 1963, front note; no further citation given there, which makes the quotation questionable).

Referring to the good old days enables the prophet to suggest by comparison that things can be made better than they now are. Salo Baron found the good-old-days idea in the Biblical prophets:

> The historian of the tenth or ninth century, the "Yahwist," as well as his successor, the "Elohist," were both men of great culture, and both were prone at least partially to accept the view, so vivid in prophetic, Nazirite and Rechabite circles, that the civilization of their own day was merely a sinful degeneration from the good old times of unspoiled primitive life (Baron 1952, 42).

Historian Norman Cohn gave us some insight into the nature and the cause of apocalyptic, millenarian thinking.

> Already [in the second century BCE, among the Jews] one can recognize the paradigm of what was to become and to remain the central phantasy of revolutionary eschatology. The world is dominated by an evil, tyrannous power of boundless destructiveness—a power moreover which is imagined not as simply human but as demonic. The tyranny of that power will become more and more outrageous, the sufferings of its victims more and more intolerable—until suddenly the hour will strike when the Saints of God are able to rise up and overthrow it. Then the Saints themselves, the chosen, holy people who hitherto have groaned under the oppressor's heel, shall in their turn inherit dominion over the whole earth. This will be the culmination of history; the Kingdom of the Saints will not only surpass in glory all previous kingdoms, it will have no successors. It was thanks to this phantasy that Jewish apocalyptic exercised, through its derivatives, such a fascination upon the discontented and frustrated of later ages—and continued to do so long after the Jews themselves had forgotten its very existence (1970, p. 21).

Jacob Talmon (1961) considers political messianism in the 18th century to be the origin of what he calls "totalitarian democracy." He deserves lengthy quotation:

> In the eighteenth century [a] peculiar state of mind . . . achieved dominance in the second part of the century. Men were gripped by the idea that the conditions, a product of faith, time and custom, in which they and their forefathers had been living, were unnatural and had all to be replaced by deliberately planned uniform patterns, which would be natural and rational (p. 3).

> [There was] intense preoccupation with the idea of virtue, which [meant] conformity to the hoped-for pattern of social harmony. They refused to envisage the conflict between liberty and virtue as inevitable. On the contrary, the inevitable equation of liberty with virtue and reason was the most cherished article of their faith. When the eighteenth-century secular religion came face to face with this conflict, the result was the great schism. Liberal democracy flinched from the spectre of force, and fell back upon the trial-and-error philosophy. Totalitarian Messianism hardened into an exclusive doctrine represented by a vanguard of the enlightened, who justified themselves in the use of coercion against those who refused to be free and virtuous (pp. 4, 5).

> Modern totalitarian democracy is a dictatorship resting on popular enthusiasm, and is thus completely different from absolute power wielded by a divine-right King, or by a usurping tyrant. In so far as it is a dictatorship based on ideology and the enthusiasm of the masses, it is the outcome . . . of the synthesis between the eighteenth-century idea of the natural order and the Rousseauist idea of popular fulfillment and self-expression. By means of this synthesis rationalism was made into a passionate faith. Rousseau's "general will" . . . became the driving force of totalitarian democracy, and the source of all its contradictions and antinomies (p. 6).

> The Right teaches the necessity of force as a permanent way of maintaining order among poor and unruly creatures, and training them to act in a manner alien to their mediocre nature. Totalitarianism of the Left, when resorting to force, does so in the conviction that force is used only in order to quicken the pace of man's progress to perfection and social harmony (p. 7).

Talmon also notes the role of the good-old-days myth:

> The strongest influence on the fathers of totalitarian democracy was that of antiquity, interpreted in their own way. Their myth of antiquity was the image of liberty equated with virtue. The citizen of Sparta or Rome was proudly free, yet a marvel of ascetic discipline. He was an equal member of the sovereign nation, and at the same time had no life or interests outside the collective tissue (p. 11).

Religious and environmental prophets through the centuries issue grim warnings of the future—hell on earth, or hell in hell—to scare people into changing how they live. The Biblical prophets sought (and seek) to change sexual behavior, moral behavior with regard to treatment of the poor, and government corruption; they aim to change both individuals and government. The environmental prophets seek to change consumption behavior—get people to use fewer natural resources, live simpler lives, refrain from policies of economic growth; they, too, seek to change both individuals and government.

Some Characteristics of Biblical and Environmental Prophecies

Biblical prophets differed in their messages, depending upon whether times were good or bad, economically and politically. In good times, they warned of punishment for present bad behavior—trouble in the future, and a fall from grace. In bad times they said that the people's misery in the present was punishment for the sins of not yet living pure ways in the past and present.

Environmental prophets now use both these messages—some saying that we are already in big trouble, others saying that we are now living high off the hog with the trouble yet to come. In both good and bad times people must "pay"—that is, suffer—for what they are doing. In bad times, people must suffer because the current misery shows that we have sinned. In objectively good times like these—clearly the best times ever for the world—we must suffer because we are enjoying too well our good fortune. (Of course if you try to tell people that things now are good, they may either deny it, or say we must pay eventually.)

Environmental prophecy does not call for us to sell our homes, stop farming, or meet on top of a mountain at a given hour to leave this place and condition for another. In this sense it is less apocalyptic than much religious prophecy, and the label "apocalyptic environmentalism" is not very appropriate.

Environmental prophecy is very much like Biblical prophecy, however, in its involvement with politics. Isaiah was typical in his concern about world and national affairs. "Isaiah dramatized his insistency [on the futility of an alliance with Egypt] by going about barefoot and naked for three years as a symbol of the fate that would overtake Egypt and its ally Ethiopia at the hands of the Assyrians" (Amsel 1994, 46).

Some have likened environmentalism to a religion. Environmental rhetoric is indeed like religion in not being the same sort of cool thinking one applies to decisions about whether or not to paint the house this year. One can see the non-business-as-usual attitude in this quote:

> Compromise will end in death. Two billion people living in poverty . . . are hostage to our greed, to our waste of energy" (agronomist Rene DuMont, quoted in *Access to Energy*, January 1990, vol. 17).

In the environmental movement there often is religious *excitement*, which is, according to Cohn (1970), the crucial element in ancient prophecy. Unlike organized religion, however, the environmental movement has no church, no service, and no priestly group. An environmentalist might say that *nature* "speaks" to us, rather than that God does.

Environmental prophecy seldom forecasts the world's ending on some specified date, unlike much religious prophecy, which derives dated predictions from particular readings of Scripture. But like religious prophecy, environmental prophecy tends to see the present age as a new phenomenon, a break with all history.

Sin plays a role in all types of prophecy. Environmentalists seize on every calamity—for example, the Exxon Valdez's fouling of Prince William Bay—as proof of our sinful ways. The Biblical prophets did so, too:

> Extreme punishment always appeared to the prophets to be impending, and finally they saw it come in the succession of national catastrophes (Baron 1952, 90).

Environmental prophets concur with religious prophets in accusing us of an excess of worldliness, and especially of enjoying the benefits of wealth. Consider, for example, the oft-heard statement that we are now in a state of "overconsumption"—that is, too much consumption. Too much relative to what? The only sensible interpretation is that the amount of present consumption is somehow sinful.

Fidel Castro at the Rio Earth Summit conference in 1992 accused the "wealthy countries" of "excessive consumption of natural resources," and he said that "the consumer societies of the developed world are totally responsible for degradation of the global environment. 'Pay the ecological debt, not the foreign debt,' he said."[4] Whether this charge is sincere or cynical on the part of Castro does not matter; what does matter is that he expects (undoubtedly correctly) that the charge will resonate with many listeners.

Why do people feel that they are in the presence of sin (and I purposely use the word "feel" rather than "think")? I interpret this as a sign of guilt that we have it so good. However, environmentalists do not charge us with sexual immorality, for the most part.

Many have said that environmentalism is socialism under another label. True, it shares with socialism the aim of putting some resources—the ocean, the forests, and the Arctic—under public ownership. But environmentalism does not call for government administration of steel mills, railroads, and retailing. It does not even talk about "distributive justice," which is not a theoretical part of socialism but has come to be its main element for many who have come to see the folly of government management of production facilities. Perhaps the issues

[4]*Washington Post*, June 14, 1992, p. A26.

become confused because many environmentalists in the United States are *also* socialists.

Both environmental and religious prophecy differ from worldly thinking in their imperviousness to counterevidence. But whereas religious prophecy is willing to be insouciant about such evidence, the environmental prophets at least pay lip service to the evidence. They simply shut their eyes to long-run trends in the evidence that contradict their prophecies, and they assert that the "theory" is to be believed rather than the evidence that conflicts with it.

In the service of their preaching, the enviro-prophets employ many appealing concepts that seem rational and even scientific, such as exponential growth and diminishing returns. But this is only a façade of pseudoscience; the prophets refuse to engage in true scientific discussion of the validity of their forecasts. In fact, the current crop of prophets will not even bet with me (winnings to charity) that their forecasts will come about. What should one think of a person who tells you to bet *your* farm that the sky will be clear tomorrow, but who is afraid to bet *his* farm or even to put out the wash on the basis of his own forecast?

The environmental prophets of today differ from Biblical prophets in that the prophets of today do not stake their lives or livelihoods on their prophecies. The Biblical prophets raised their hands against their rulers, usually at great risk; today's crop is part of the establishment.

Organized environmental lobby groups are very much in the Biblical tradition. In Biblical days there were groups of "disciples of the prophets" whose "number [was] at times in the hundreds" (Amsel 1994, 1156). And the head prophets then, as now, were sometimes well paid—"as much as 40 camels bearing the treasures of Aram" (Amsel 1994, 1157)—just as the leaders of the environmental movement (especially in Washington) are paid comparably with heads of other lobby groups.

Are All Prophets False?

If the concept of prophecy has any meaning, it must refer to some form of supposed knowledge different from the ordinary practices of science and rational decisionmaking that we may call "worldly thinking." But if the process of science is a "true" method, then any system of thinking and forecasting that departs from the practices of worldly thinking must *by definition* be a false method, though in individual cases the prophets' forecasts may turn out to be correct.

To my knowledge prophets have never claimed that theirs is a reliable process of worldly knowledge, but only that particular forecasts are correct. Hence it is reasonable to conclude that the method of prophecy is false with respect to worldly forecasts.[5]

The question of the truth or falsity of prophets has the nature of a Russell-and-Whitehead paradox, of the all-Cretans-are-liars sort. Salo Baron has written somewhere that those who forecast good tidings are automatically regarded as false prophets. This means that if good times really are in the offing, a true prophecy implies that the maker is a false prophet. To avoid being a false prophet, the person must make the false prophecy that the tidings are bad. So unless the tidings always are bad, prophecy can hardly meet the test of ordinary science and ordinary truth.

The environmentalism movement takes this notion to the limit. It gives voice to false prophecies uttered by false prophets. These sayers of doom forecast that because our ways are evil, our environment will deteriorate—that it will become less healthy and less clean. And on the basis of these prophecies, many citizens and politicians then call for strong remedies that could cause more damage than good.

How may one judge the forecasts? We cannot be certain what the future holds. But we do know the results of similar prophecies in the past. The doomsaying prophets of the 1960s and 1970s—the same people who are the false prophets of today—were wrong across the board. And prophets in previous eras also have a very poor record.

[5]I apply the label "false" not as logic-chopping but as a way of distinguishing one method from another for practical purposes.

We also have irrefutable evidence that the assertions of fact on which the doomsters supposedly base their prophecies are false. They say that our air and our water have been getting dirtier in recent decades, and that natural resources have been becoming more scarce in past decades and centuries. These propositions are demonstrably, indubitably, wrong.[6] There is every reason to believe that the current environmental prophecy of doom is wrong.

Given that the supposed evidence for the prophecies is false, it is reasonable to say that the prophecies built upon them are false, even if by chance one or more of them might come about. And if the prophecy is indeed wrong, the prophets are dangerous. That's my message in a nutshell.

In the Bible, God agrees that a prophet's forecasting record is all-important. First God promises that prophecies will not be made frivolously:

> A prophet will the Lord thy God raise up unto thee, from the midst of thee, of thy brethren, like unto me; unto him ye shall hearken; . . . I will put My words in his mouth, and he shall speak unto them all that I shall command him.

God also gives the prophet plenty of authority:

> And it shall come to pass, that whosoever will not hearken unto My words which he shall speak in My name, I will require it of him.

But God also makes it clear that prophetic performance is crucial, and there should be no excusing those who prophesy erroneously:

[6]Every agricultural economist knows that the world's population has been eating ever better since World War II. Every resource economist knows that all natural resources have been getting more available rather than more scarce, as shown by their falling prices over the decades and centuries. Every demographer knows that the death rate has been falling all over the world—life expectancy almost tripling in the rich countries in the past two centuries and almost doubling in the poor countries in just the past four decades. And even the Environmental Protection Agency, which has a stake in continuing problems, acknowledges that our air and our water have been getting cleaner rather than dirtier in the past few decades.

> But the prophet, that shall speak a word presumptuously in My name, which I have not commanded him to speak, or that shall speak in the name of other gods, that same prophet shall die (Deut. 18:15-22).[7]

The newspapers and television take a different view of prophecy about the environment, resources, and population growth. False performance over the past 30 years at least has been rewarded with ever-greater attention by the press. And correct forecasting has not brought greater attention and respect by the press; mostly, it has brought obloquy.

Why Do People Believe Prophecies That Are So Wrong?

Why do people hold beliefs contrary to all of human experience for thousands of years, such as that raw materials will become more scarce and their prices will rise? One can say that people are under the sway of a convincing theory—that the supply ultimately is limited. But why do people hold onto such a theory even when, as they admit, it is contradicted by the evidence?

Again and again over the past quarter century, after people see the data showing that all trends pertaining to human welfare have been improving rather than deteriorating—health, wealth, education, leisure, availability of natural resources, cleanliness of our air and water, you name it—the question arises: Why, then, do our media and our political leaders tell us the opposite—that life is more dangerous, our planet is "plundered" and "in crisis," we are running out of resources, pollution is increasing—that is, that things are getting worse when they are really getting better? Why do we hear that there is need to "save the planet"?

There are many strands in an explanation of why so many people hold these beliefs against all available evidence. They include *institutional* and *economic* influences such as advocacy groups making statements that are proven money-getters in fundraising letters, and scientists "discovering" problems that

[7]Thanks to Kathy Rochelle, who brought these verses to my attention.

will elicit funding for research to mitigate the problem. There are also *intellectual-cognitive* causes of error, a complex subject that I treat at length elsewhere (1990, selection 52); among our many human limitations on clear thinking I'll mention only (1) the zero-sum mentality; (2) the belief in finiteness as a starting point for reasoning, which leads to the belief that eventually there *must* be exhaustion; (3) lack of understanding of the process of building wealth and resources; and (4) the belief that even if they are wrong in their admonitions to people to change their behavior, the results cannot be harmful because the ensuing behavior will be more moral, whether that be more modest dress for the women that the Biblical prophets scolded or the less resource-using economic activities that environmental prophets advocate; this logic of there being no harm even if one is wrong is embodied in Paul Ehrlich's version of Pascal's wager, which calls for population reduction, fewer people having no negative effect of any kind in his value system.

False prophecy is certainly not the only example of ingrained belief that runs against the available evidence. One might also ask why people hold beliefs about sports behavior and the stock market even after they have been informed that the beliefs have been demonstrated to be false? Why do people believe in ESP in connection with dreams about a death or a chance meeting that actually took place later? Yes, one can argue that the probability theory of coincidences is by no means obvious. But even after the holders of such beliefs are presented with such "rational" explanations of the coincidence, they often continue to hold onto the belief in ESP. And how can believers in a doomsday continue to believe after the date has passed?

But for the prophets themselves I think there is a third and dominating element that defies simple cost-benefit rational explanation: There seems to be a built-in, age-old psychological propensity to hark back to the "good old days" and warn of impending doom in some people that we might call the urge to prophesy; this need may be connected with many emotions.

One sees the good-old-days syndrome even in the most level-headed of people, even in the fields they know well.[8] David Hume (1987) said:

> The humour of blaming the present, and admiring the past, is strongly rooted in human nature, and has an influence even on persons endued with the profoundest judgment and most extensive learning (p. 464).

This suggests that we should not seek a "rational" explanation for prophecies of doom—that is, an explanation that "makes sense" in terms of the elements of the situation of which the prophet is aware—at least at this stage of knowledge of human psychology. After all, can we explain why so many people—including this writer—"knock on wood" when someone comments that things are going well, for fear that the good fortune will not continue? Can we explain why many people wear amulets of shapes that are thought to keep away

[8]Consider the case of the great British constitutional lawyer A. V. Dicey (1915):

DECLINE IN REVERENCE FOR RULE OF LAW

> The ancient veneration for the rule of law has in England suffered during the last thirty years a marked decline. The truth of this assertion is proved by actual legislation, by the existence among some classes of a certain distrust both of the law and of the judges, and by a marked tendency towards the use of lawless methods for the attainment of social or political ends (p. lv).
>
> Within the last thirty years, however, there has grown up in England, and indeed in many other civilised countries, a new doctrine as to lawlessness. This novel phenomenon, which perplexes moralists and statesmen, is that large classes of otherwise respectable persons now hold the belief and act on the conviction that it is not only allowable, but even highly praiseworthy, to break the law of the land if the law-breaker is pursuing some end which to him or to her seems to be just and desirable. This view is not confined to any one class. Many of the English clergy (a class of men well entitled to respect) have themselves shown no great hesitation in thwarting and breaking laws which they held to be opposed to the law of the Church. Passive resisters do not scruple to resist taxes imposed for some object which they condemn. Conscientious objectors are doing a good deal to render ineffective the vaccination laws. The militant suffragettes glorify lawlessness; the nobleness of their aim justifies in their eyes the hopeless and perverse illegality of the means by which they hope to obtain votes for women.
>
> Whence arises this zeal for lawlessness? (p. lix).

the evil eye? Or patronize fortune tellers? Or consult the horoscope section of the newspaper, one of the best-read features in the daily press? Yes, one can say that people "feel a need" to make sense of a world which is unknown in its uncertainty, but why do people feel such a need?

If we observe the same sort of thought and action in all times and places, it is reasonable to assume that there is something instinctive about it. And we have seen earlier that prophecies of doom have always been with us. The best we can do, I think, is to agree with David Hume and Friedrich Hayek that during the course of physical and social evolution we have acquired many instincts that do not accord with what we call reason, though the instinctual actions may have had some usefulness in some situations along the way. One such instinct is the propensity to prophesy doom.

The Importance of False Prophecy

Doomsaying may seem harmless in the short run. Many believe that incorrect prophecy can have a positive effect in forestalling complacency. But false prophecy can cause huge long-run disaster.

One possible evil of false prophecy is social disruption. Examples include the Shabtai Zevi episode that devastated European Jewry in the Middle Ages, and the more recent Heaven's Gate and Branch Davidians (David Koresh) events.

Another possible evil is the totalitarian messianism that Talmon wrote about. The Salem witchcraft trials are likely to have been the result of Michael Wigglesworth's *The Day of Doom* (Starkey 1949), an instant best-seller published in 1662; one copy was quickly sold for every 20 people in New England; that was followed by four more American editions and eight in London (Murdock 1966). The parallel between *Day of Doom* and Rachel Carson's *Silent Spring* (1962) has been mentioned by several scholars.

False prophecy can also exact resource losses. The false crises of oil and other natural resources in the 1970s were very costly for Japan, the U.S. airplane makers, and many others.

For individuals, doomsday predictions result in families quitting their jobs and selling their household possessions, as happened to the Davidians in 1959 when founder Victor Houteff's widow proclaimed that on Easter the Kingdom of God would arrive.[9] In former times, farmers stopped tending their fields. The owner of 38 radio stations said that after earthquakes and other troubles, Jesus Christ would return in glory between September 15 and September 27, 1994; between his radio stations and the 40,000 copies of his book, who knows how many people disrupted their lives?[10]

What Is the Future of Prophecy?

Will there be less or more of religious and environmental prophecy in the future? There is little basis for a prediction.

Maybe we should ask what constrains the amount of false prophecy from increasing wildly? I believe that only the demands of the everyday necessity to make a living constrain the flights of fantasy and the excitement of prophecies made and believed. And as society becomes richer, fewer people and groups are prevented by this necessity from indulging themselves in these emotional orgies.

Maybe only some big challenge—hopefully something positive, such as exploration of space, though it could be from a catastrophe of war or disease—can act as a bulwark. But at this point of speculation, the discussion must come to a close.

[9] *Newsweek*, March 15, 1993, p. 57.

[10] *Washington Post*, March 4, 1993, p. A1.

Conclusion

With reasonable surety one can expect that the material conditions of life will continue to get better for most people, in most countries, most of the time, indefinitely. Within a century or two, all nations and most of humanity will be at or above today's Western living standards. The basis for this forecast is the set of trends discussed in my other books and summarized in the introduction to this one.

Whatever nature has spontaneously produced that we use—food, oil, diamonds—humankind now can also produce by design, and faster than nature. An expectancy of health and a standard of living higher than that which any prince or princess enjoyed 200 years ago now is the birthright of every middle-class and working-class person in developed countries, and of most people in poverty as well. What is still to come is to bring these material gains to all groups of humanity. That may take half a century or a century. Yet that benign outcome may be predicted with high likelihood. This is a happy vision, indeed.

Why should you believe this forecast rather than the forecasts made by the doomsayers?

- This "side" has been right across the board in the forecasts we have made in the past few decades, whereas the doomsayers have been wrong across the board.
- Throughout the long sweep of history, forecasts of resource scarcity have always been heard, and—just as now—the doomsayers have always claimed that the past was no guide to the future because they stood at a turning point in history. But the turning point forecasts have been wrong; there have been ups and downs, but no permanent reversals. In every period those who would have bet on

improvement rather than deterioration in fundamental aspects of material life—such as the availability of natural resources—would usually have been right.

- I'll bet my reputation and my money on these forecasts (if I win, the money goes to pay for research), whereas the doomsayers back off from putting their money where their mouths are; they refuse to put either their cash or their names on the line to back what they say. Indeed, the most famous of the doomsayers was burned badly when in 1980 his group actually did wager on some of his forecasts.

The doomsters' unwillingness to make wager commitments should call into question whether they really believe the dire forecasts that they make, or they just make statements they don't believe to scare the public and mobilize the government to do their will.

But I make another and darker forecast, though admittedly based on much less solid evidence: Many people will continue to *think and say* that the material conditions of life are getting *worse* even though the conditions actually are getting better. This assessment will become more cheerful only when (or if) humanity invents or evolves or stumbles into an invigorating set of new challenges that will capture people's imaginations and hearts and wills and replace the intergroup political struggles that now increasingly supplant the struggle against nature for a better material life.

Why should people's perceptions of the trends be the opposite of the facts? If we were all clear-minded and powerful thinkers, false bad news would have little effect; our wonderfully functioning minds would spot and filter out the unsound information. But instead we are all of us endowed with very imperfect and biased thinking mechanisms. How much of our frail and idiosyncratic thinking is born into us and how much we acquire along the way is not now known. But it is sure that by the time we reach adulthood we are full of intellectual frailties such as:

- our willingness to rely on and believe a single dramatic account on television, or a single year's statistic, that runs counter to a large quantity of data collected over many years that run in the opposite direction;

- our propensity to believe (especially as we get older) that the "old days" were better physically and morally than our own times; and
- our inclination to believe that the quantity of expression of one point of view given space in the newspapers and time on television, and the number of organizations urging that view—for example, that our air and water are getting dirtier rather than cleaner—is a good guide to the likely validity of that view.

Given that we will not all turn into clear-minded objective thinkers tomorrow, we must therefore inquire into the sources of our beliefs. Whatever else is also true, the future for the correct perception of these trends looks bleak because of their inaccurate portrayal in the press. One cause of the media's inaccuracy is that the techniques that journalists use so well to cover fires and local politics do not work well for matters that go beyond first-hand observation. This includes scientific matters, as illustrated nowadays by environmental questions. And the bad-news bias in journalism turns every story negative even if the underlying facts are positive. This leads the public to think that conditions in general are getting worse. The press then reports this as pessimism. This self-reinforcing process could have increasingly dire effects upon the public mood.

Does the Market for News Fail Us?

Is this a case of what some economists call "market failure"? That is, are the economic forces in this case insufficient to provide what one could expect a well-functioning market to provide? I do not think so. It is not reasonable to expect free markets to always bring us what is good for society and individuals within any foreseeable period, if ever. The conditions in some markets inevitably produce a certain amount of fraud; an example is the encyclopedia industry, where the salesperson expects never to see the customer again after a purchase is made, and therefore there is no incentive to keep the customer's good will—no market sanction—for providing misinformation about the product and its use. Other industries inevitably provide goods that are bad for many people; hard drugs are an easy example. And special government intervention beyond the enforcement of the basic laws governing markets may do more harm than good in such cases.

It is entirely to be expected that some journalists, just like persons in any other occupation, will bend the ethics of their occupation for personal gain and to advance their own ideological agendas. The problem in journalism is that the forces to check this very human propensity are too weak. I do not have any easy remedies to suggest. I hope that the personal integrity of journalists and media owners will lead them to hew to high standards. This is not at all an unrealistic hope, or one that is at odds with the notion of free enterprise; anyone who thinks that business is by nature dog-eat-dog with no holds barred other than what will get you thrown in jail entirely misunderstands the nature of a free-enterprise economy and society. On the other hand, anyone who expects all persons to hew to such high standards, or for the market always to enforce high standards in the long and short run, is charmingly naive.

The more competitors in the news marketplace the better. The greater the number of firms competing, the more it will be profitable for at least a few of the many competitors to show how others are scandalously misinforming the public; muckraking the false muckrakers can be exciting news, too. False bad news will also draw the attention and energies of true do-gooders who will operate think tanks and other not-for-profit private activities that turn the searchlight of truth onto the Willie Suttons of the press, and squeeze them a bit.

It may well be, however, that there will always be structurally based incentives to produce a lot of false bad news. Only the future of the news media will reveal the answer to that question, and I do not know of any studies of the trends that would provide a solid basis for forecast. As noted in the introduction, there are few obvious ways to improve the structure in a free society. Getting rid of the long-time gentleman's agreement for the media not to criticize each other may be the most promising alteration in the structure, and we may already be seeing some change in that direction.

References

Amsel, Nachum. 1994. *The Jewish Encyclopedia of Moral and Ethical Issues.* Northvale, N.J.: J. Aronson.

Baron, Salo Wittmayer. 1952. *A Social and Religious History of the Jews.* New York: Columbia University Press.

Baumol, William J., and Wallace E. Oates. 1995. "Long-Run Trends in Environmental Quality." Chapter 43 in *The State of Humanity.* ed. Julian L. Simon. Cambridge, Mass.: Blackwell.

Beckmann, Petr. 1978. *What Attracts Intellectuals to Socialism?* Boulder, Colo.: Golem Press.

Bennett, William J. 1993. *The Index of Leading Cultural Indicators.* Vol. 1. Washington, D.C.: Empower America, the Heritage Foundation, Free Congress Foundation.

Bernal, J. D. 1953/1970. *Science and Industry in the Nineteenth Century.* Reprinted, Bloomington: Indiana University Press.

Bloom, David E. 1995. "International Public Opinion on the Environment." *Science* 269 (July 21): 354–8.

Carson, Rachel. 1962. *Silent Spring.* Boston: Houghton Mifflin.

Clark, Mary E. 1989. Letter to editor. *Science* 246 (October 6): 10.

Cohn, Norman. 1970. *The Pursuit of the Millennium.* London: Granada Publishing.

Cook, Kenneth A. 1981. "The National Agricultural Lands Study Goes out with a Bang." *Journal of Soil and Water Conservation* (March/April): 92.

Davis, Shelley. 1984. "Roots under Water." *Washington Post,* April 15, pp. D1, D4.

Dicey, A. V. 1915. *Introduction to the Study of the Law of the Constitution.* 8th ed. Reprint, Indianapolis, Ind.: Liberty Classics, 1984.

Dunlap, Riley E., and Rik Scarce. 1991. "The Polls—Poll Trends." *Public Opinion Quarterly* 55: 651–72.

Ehrlich, Paul R. 1968. *The Population Bomb.* New York: Ballantine.

Ehrlich, Paul R., and Anne H. Ehrlich. 1990. *The Population Explosion.* New York: Simon and Schuster.

Field, Roger. 1988. "Old Macdonald Had a Factory." *Discover* (December): 46–51.

Fischel, William A. 1985. *The Economics of Zoning Laws.* Baltimore, Md.: The Johns Hopkins University Press.

Frieden, Bernard J. 1979. *The Environmental Protection Hustle.* Cambridge, Mass.: M.I.T. Press.

Gilbert, Dennis A. 1988. *Compendium of American Public Opinion*. New York: Facts on File.

Gore, Albert, Jr. 1992. *Earth in the Balance: Ecology and the Human Spirit*. Boston, Mass.: Houghton Mifflin.

Harris, Louis, and Associates. 1980. Report to Marsh and McClennan, "Risk in a Complex Society."

Harrison, Gordon. 1978. *Mosquitos, Malaria, and Man*. New York: E.P. Dutton.

Hayek, Friedrich. 1952. *The Counter-Revolution of Science*. Glencoe, Ill.: The Free Press.

Hochschild, Jennifer L. 1995. *Facing Up to the American Dream*. Princeton, N.J.: Princeton University Press.

Homer, Sidney. 1963. *A History of Interest Rates*. New Brunswick, N.J.: Rutgers University Press.

Hume, David. 1987. *Essays, Moral, Political, and Literary*. ed. Eugene F. Miller. Indianapolis, Ind.: Liberty Classics.

Kristol, Irving. 1978. *Two Cheers for Capitalism*. New York: Basic Books.

Krug, Edward C. 1992. "The Great Acid Rain Flimflam." In *Rational Readings on Environmental Concerns*, pp. 35–43. ed. Jay Lehr. New York: Van Nostrand.

Kuznets, Simon. 1965. "Demographic Aspects of Modern Economic Growth." Paper presented at the World Population Conference, Belgrade, Yugoslavia, September.

Lewis, H. W. 1992. *Technological Risk*. New York: Norton.

Lindzen, Richard S. 1992. "Global Warming: The Origin and Nature of Alleged Scientific Consensus." Cambridge, Mass.: M.I.T.

Lipset, Seymour Martin, and William Schneider. 1987. *The Confidence Gap—Business, Labor, and Government in the Public Mind*. Rev. ed. Baltimore, Md.: The Johns Hopkins University Press.

Maduro, Rogelio A., and Ralf Schauerhammer. 1992. *The Holes in the Ozone Scare*. Washington, D.C.: 21st Century Science Associates.

Malthus, Thomas R. 1803. *An Essay on the Principle of Population, or a View of Its Past and Present Effects on Human Happiness*. 2d ed. London: J. Johnson.

McBride, Stewart. 1980. "Doomsday Postponed." *Christian Science Monitor*, (August 26): 810–11.

Meadows, Donella H., Dennis L. Meadows, Jorgen Randers, and William W. Behrens III. 1972. *The Limits to Growth*. New York: Universe Books.

Merida, Kevin. 1995. "Worry, Frustration Build for Many in Black Middle Class." *Washington Post*, October 10, 1995.

Mills, Stephanie. 1991. "Population—Red-Hot Realities for a Finite Planet." *Garbage*, May-June 1991, pp. 46–51.

Mises, Ludwig Von. 1972. *The Anti-Capitalist Mentality*. South Holland, Ill.: Libertarian Press.

Murdock, Kenneth B. 1966. Introduction to *The Day of Doom*, by Michael Wigglesworth. New York: Russell and Russell.

National Research Council, Committee on Population, and Working Group on Population Growth and Economic Development. 1986. *Population Growth and Economic Development: Policy Questions*. Washington, D.C.: National Academy Press.

Nisbet, Robert, and Lee Ross. 1980. *Human Inference*. Englewood Cliffs, N.J.: Prentice Hall.

Norman, Geoffrey. 1988. "The Hustle and Hypocrisy of Andrew Young's Atlanta." *The American Spectator*, June 1988, p. 24.

O'Neill, Catherine. 1991. "Seek a Solution to Pollution." *Washington Post*, August 6, Health Section, p. 18.

Osborn, Fairfield. 1953. *Limits of the Earth*. Boston, Mass.: Little Brown.

Petersen, William. 1976. "An 'Optimist's' Pessimistic View of the Food Situation." *Social Science Quarterly*: Vol. 57, p. 365.

Petersen, William. 1977. "On a Writer Unable to Read." *Social Science Quarterly*: Vol. 58, p. 244.

Plous, Scott. 1993. *The Psychology of Judgment and Decision Making*. New York: McGraw-Hill.

Popper, Karl R. 1966. *The Open Society and Its Enemies*. 5th ed. Princeton, N.J.: Princeton University Press.

Postman, Neal. 1992. *The Surrender of Culture to Technology*. New York: Knopf.

Price, Daniel O., ed. 1967. *The 99th Hour*. Chapel Hill: University of North Carolina Press.

Rathje, William L., and Cullen Murphy. 1992. *Rubbish: The Archaeology of Garbage*. p. 162. New York: HarperCollins Publishers.

Raven, Peter H. 1988. "The Cause and Impact of Deforestation." In *Earth '88—Changing Geographic Perspectives*, pp. 212–29. ed. H.J. de Blij. Washington, D.C.: National Geographic Society.

Reston, James. 1991. *Deadline*. New York: Random House.

Rifkin, Jeremy. 1989. *Entropy: Into the Greenhouse World*. Rev. ed. New York: Bantam.

Rosenberg, Nathan. 1972. *Technology and American Economic Growth*. New York: Harper.

Sax, Karl. 1960. *Standing Room Only: The World's Exploding Population*. 2d ed. Boston, Mass.: Beacon Press.

Shapiro, Leo J. 1991. "Leveraging Our Luck." *Contingencies* (Nov./Dec.): 26–32.

Sherman, Stratford P. 1990. "How High School Kids See the 1990s." *Fortune* (March 26): 226.

Simon, Julian L. 1980. "Resources, Population, Environment: An Oversupply of False Bad News." *Science* 208 (June 27): 1431-7.

________. 1981/1996. *The Ultimate Resource*. 2d ed. Princeton, N.J.: Princeton University Press, 1996.

________. 1982. "Are We Losing Our Farmland?" *The Public Interest* 67 (Spring): 49–62.

———. 1990. *Population Matters*. New Brunswick, N.J.: Transaction.
———. 1993. *Good Mood: The New Psychology of Overcoming Depression*. Chicago: Open Court.
Simon, Rita J. 1985. *Public Opinion and the Immigrant*. Lexington, Mass.: Lexington Books.
Sirico, Robert A. 1995. "The False Gods of Earth Day." *Wall Street Journal*, April 22, 1995, editorial page.
Sommer, Adrian. 1976. "Attempt at an Assessment of the World's Tropical Mist Forests." *Unasylva* 28(112–113): 5–24.
Starkey, Marion L. 1949. *The Devil in Massachusetts*. New York: Knopf.
Synthesis Group. 1991. *America at the Threshold: Report of the Synthesis Group on America's Space Exploration Initiative*. Washington, D.C.: Government Printing Office.
Talmon, J. L. 1961. *The Origins of Totalitarian Democracy*. New York: Praeger.
Vogt, William. 1948. *Road to Survival*. New York: Sloane.
Wallis, W. Allen, and Harry V. Roberts. 1956. *Statistics: A New Approach*. Glencoe, Ill.: Free Press.
Weiss, Julian. 1983. "The Farmland—Urbanization Story: A Study of Media Coverage." Washington: Typescript.
Wertham, Frederic. 1969. *A Sign for Cain: An Exploration of Human Violence*. New York: Warner Paperback.
Whelan, Elizabeth M. 1993. *Toxic Terror: The Truth Behind the Cancer Scare*. Buffalo, N.Y.: Prometheus Books.
Wigglesworth, Michael. 1662. *The Day of Doom*. Ed. Kenneth Murdock. New York: Russell & Russell.

Index

Agriculture
advances in, 8
false information about disappearing farmland, 17–29
American Farmland Trust, 24, 26–27
Amsel, Nachum, 112, 114
Assessments
bias in, 93–95
of the good old days, 96–97, 103–5, 108–11
of local opposed to national, 94–99
media influence on, 97–99
people's inaccurate, 124–25
Audubon Society, 9

Baron, Salo, 109, 115
Baumol, William J., 59
Beckmann, Petr, 40
Beliefs
about evolution of technical knowledge, 36–37
about the past and the future, 103–5, 107–21
in false prophecies, 117–20
held against available evidence, 1–3, 117
of public about environment, resources, and population, 9–14, 108–11
related to centralized government control, 39–40
of reports of the environment, 27–28
how sources influence, 101–3
Bennett, William, 83–86
Bergland, Bob, 21
Bernal, J. D., 36–37
Bias
of journalists, 79–80
in people's assessment of situations, 93–95
Biological research
ahistorical character of, 58–59
assertions based on, 57–68
Biologists
assertions about resources and environment, 57–68
impervious to new information, 67
predictions about resources, 66–68
research techniques, 59–68
view of human ecology, 62–68
warnings of resource scarcity and population growth, 55–58
Block, John, 23
Bloom, David E., 9
Borrow pits, 41
Brewer, Michael, 21, 23
Broder, David, 97
Brown, Lester, 28, 79, 89
Burke, Edmund, 45

Cannon, Walter, 68
Carson, Rachel, 69, 88, 120
Castle, Emery, 20
Castro, Fidel, 113
Cato Institute, 46
Clark, Mary E., 56
Cohn, Norman, 109, 112
Competitive Enterprise Institute, 46
Conable, Barber, 64–65
Consensus
  of economists about resources, 31–33
  of scientists about environment, resources, and population, 1
Cook, Kenneth A., 22–23

Data
  biologists' uses of, 56–62
  combining apples and oranges, 85
  ignoring and manipulating accurate, 73–75, 83–92
Data sources
  Census Bureau, 22–23
  on media handling of disappearing farmland story, 27
  National Agricultural Lands Survey, 18–27
  USDA Economic Research Service, 20
  USDA Soil Conservation Service, 20–24
  used to provide false information, 84–92
  See also Polls; Surveys.
Davis, Bernard, 57
Davis, Shelley, 43
Day of Doom (Wigglesworth), 120
Department of Agriculture
  data of Economic Research Service, 20
  data of Soil Conservation Service, 19–23
  retraction of false information, 24–25
  as source of farmland scarcity information, 18–27
Depression, psychological, 101
Diapers, disposable, 9
Dicey, A. V., 119n8
Doomsayers
  biologists as, 55–58
  forecasts and prophecies of, 11–12, 107–8, 123–24
  lack of historical perspective, 34–35
  wanting more government intervention, 38

Earth in the Balance (Gore), 83, 87
Easterbrook, Gregg, 25
Ecologists
  arguments with economists, 68–71
  conception of man and nature, 68–69
Ecology, human and animal, 62–65
Economics
  criticism of analysis using, 55–56, 62–64
  ecologists' arguments with economists, 68–71
  effect of population growth on development, 8
  of free lunch, 34

of natural resources, 31
of population, 35–37, 48–49
rejection by biologists, 62–68
Ehrlich, Paul, 28, 49, 55, 56, 59, 62–64, 66, 89
Encyclopedia industry, 125
Energy concerns, 10–11
Environment
Americans' concern about, 95
cleaner, 7
differing views of ecologists and economists, 68–71
false bad news about, 1–3, 15–16
idea of crisis related, 10–16
media reporting on, 73–81
media's wrong information about, 73
organizations telling truth about, 50–53
prophecies and rhetoric about, 111–14
public's false beliefs about, 9–12
See also Pollution.
Environmentalists
campaigns of, 49–50
concern related to externalities of economic activity, 41–42
false prophecies of, 114–17
Environmental Protection Agency, 7, 10
Essay on Population (Malthus), 66

Facing Up to the American Dream (Hochschild), 91–92
Farmland, disappearing, 17–29
Farmland Protection Policy Act (1981), 27
Field, Roger, 48
Fischel, William, 19, 28
Food and Drugs Act (1906), 80
Food productivity increase, 28
Forecasts
about future standard of living, 123–24
of doom based on good old days, 108–11
of doomsayers, 10–11, 107–8, 123–24
of resource scarcity, 123–24
Frey, H. Thomas, 20–21

Garbage disposal, 41
Gardner, Bruce, 87
Godwin, William, 38
Goodland, Robert, 64
Gore, Albert, Jr., 83, 87–90
Government
arguments for and against intervention by, 38
belief in centralized control by, 39–40
information about disappearing farmland, 17–29
Grass, greener, 95
Gray, Robert, 23–24, 27

Handler, Philip, 2, 88
Hardin, Garrett, 28, 51, 56–57, 61, 64, 67
Harris, Louis, and Associates, 13–14, 16
Harrison, Gordon, 88
Hart, John Fraser, 20, 47
Harwood, Richard, 78
Hayek, Friedrich A. von, 39–40, 120
Heady, Earl, 87

Heidelberg Declaration, 49
Hochschild, Jennifer, 91–92
Homer, Sidney, 109
Human nature, 38
Hume, David, 119, 120

The Index of Leading Cultural Indicators (Bennett), 83–86
Information
  biologists' incorrect, 58–59, 66–67
  of media about population, 73
  media dissemination of false, 73–91
  media sources of, 93–99
  needed organization for truthful, 50–53
  related to acid rain and dioxin, 46–47
  statistics used to provide false, 83–92
  See also Data; Data sources; Media, the; News; Statistics.

Journalists
  built-in bias, 79–80
  defense of standard practices, 79–80
  use of substitutes for scientific research, 78–80
  See also Media, the.

Kaczynski, Theodore J, 108
Kahn, Herman, 47
Knowledge
  fragmentation of, 46
  increased population creates new, 70–71
  more people create more, 36–37
  personal, 93–105
  See also Assessments; Beliefs; Bias
Koresh, David, 107
Kristol, Irving, 107–8
Krug, Edward, 46, 47
Kuznets, Simon, 32, 45n1

LaPiere, Richard, 98
Lee, Linda, 24
Lewis, H. W., 48
Life expectancy, 8
Limits of the Earth (Osborn), 55
The Limits to Growth (Meadows et al.), 55
Lindzen, Richard S., 46, 47
Lipset, Seymour Martin, 93
Lovins, Amory, 70
Lutrell, Clifford, 19

McBride, Stewart, 59
McLaren, Dale, 22
Malthus, Thomas, 31–32, 38, 65, 66
Meadows, Donella H., 55
Media, the
  beliefs about resources, population, and the environment, 28–29
  incentives in false reporting, 125–27
  as source of information, 93–99
  wrong information about population and environment, 73
Medical advances, 8, 47
Merida, Kevin, 90–91
Mills, Stephanie, 56

Mortality rates
  black infants, 104
  decrease in rich and poor countries, 7–8
Mosquitoes, Malaria, and Man (Harrison), 88
Murdock, Kenneth B., 120

National Academy of Sciences, National Research Council, 8, 31, 66
National Agricultural Lands Study (NALS), 18–27
National Center for Policy Analysis, 46
News
  actions against false bad news, 45–46
  bad news about disappearing farmland, 27–29
  competition for, 126–27
  false bad news, 1–5, 9–14
  improving quality of, 80–81
  incentives to produce false, 2–3
The 99th Hour (Price), 65
Nisbet, Robert, 61
Norman, Geoffrey, 74
Nuclear power debate, 42–43

Oakes, John, 35
Oates, Wallace, 59
Osborn, Fairfield, 55

Petersen, William, 67
Plous, Scott, 98
Political Economy Research Center, 46
Polls
  about disposable diapers, 9
  intepretation of survey of blacks and whites, 90–92
  positive and negative responses, 99
  responses to questions about federal budget, 97–99
  scientific polling methods, 93
Pollution
  differing perceptions of, 41–42
  perception of trends related to, 14
  public concerns about, 14
Popper, Karl, 41
Population
  better fed, 7
  correlation of size, knowledge and productivity, 36–37
  false bad news about, 1–3
  in finite or closed Malthusian system, 31–33
  impact of large, 37–39
  increase in world's, 7–8
  in infinite or open system, 31–33
  media's erroneous information about, 73
  public's false beliefs about, 9–14
Population Bomb (Ehrlich), 55
Population growth
  benefit of, 18
  biologists' warnings of, 55–58
  effect on economic development, 8
  increased demand for goods and services with, 36–37, 71
  Lewis's conception, 48–49
  long- and short-run trends in, 35–36

pre- and post-eighteenth century, 7–8
telling truth about, 50–53
Postman, Neal, 108
Press, the
beliefs about the environment, 27–29
responsibility for accuracy, 92
role in disappearing farmland news, 25–26
See also Media, the.
Price, Daniel O., 65
Prices
farmland, 28
of natural resources, 34
Productivity, 36–37
Prophecies
biologists as doomsayers, 55–58
environmental and religious, 111–14
false, 107–21
of 1970s about population and the environment, 28–29
role of sin in, 112–13
true and false, 114–17, 120–21
Public, the
false beliefs about environment, resources, and population, 9–14
messages of doomsayers to, 45–46
polls misleading, 77–78

Raspberry, Wiliam, 78
Raven, Peter, 49, 56
Representativeness concept, 60
Research
biological, 58–62
biologists speaking outside fields of their, 62–68
The Resourceful Earth (Kahn and Simon), 47
Resources, natural
biologists' doomsayer arguments about, 55–58
biologists' predictions about, 66–68
consensus of economists, 31
false bad news about, 1–3
in finite or closed Malthusian system, 31–33
forecasts of scarcity, 55–58, 123–25
historical real prices of, 34
hydroponic cultivation of crops, 43–44
in infinite or open system, 31–33
media reporting on, 73–91
more available, 7
organizations telling truth about, 50–53
public's false beliefs about, 9–14
Reston, James, 35, 78
Rhetoric, environmental, 112–13
Rifkin, Jeremy, 108
Ringle, Ken, 74
Rio Earth Summit (1992), 46
Road to Survival (Vogt), 55
Roberts, Harry, 83–84
Rosenberg, Nathan, 32
Ross, Lee, 61

Salk, Jonas, 60
Sanitation advances, 8
Sax, Karl, 55

Schneider, William, 93
Schultz, Theodore, 87
Scientific methods
  circumstances for using, 75–77
  failure to use, 77–78
  impact of not using, 77–78
  journalists' substitutes for, 77–80
Scientists
  positions on environmental issues, 47–49
  spotting false scares, 46–50
  See also Biologists.
Silent Spring (Carson), 69, 88, 120
Simon, Julian, 2, 28, 36, 94, 102
Simon, Rita, 95
Sirico, Robert A., 11
Sommer, Adrian, 60
Stamps, self-sticking, 10
Standard of living
  biologists' warnings about declining, 55–56
  with more people, 36–37
  present and future, 123–24
  trends in deteriorating, 13–14
Standing Room Only (Sax), 55
Starkey, Marion L., 120
Statistics
  Bennett's misuse of, 83–86
  biologists' distrust of aggregate, 61
  biologists' use of, 59–61
  Gore's misuse of, 87–90
  Hardin's contempt for analyses using, 56–58
  policy issues requiring analysis using, 73–91
  sampling, 73–77
  used to provide false information, 83–92
  See also Data; Data sources; Information; News
Statistics: A New Approach (Wallis and Roberts), 83
Substitutes for science, 78–80
Sudman, Seymour, 21
The Surrender of Culture to Technology (Postman), 108
Surveys
  about the environment, 12–16
  National Agricultural Lands Survey, 18–29
Swanson, Earl, 87

Talmon, Jacob, 110–11, 120
Tazieff, Haroun, 46–47
Technological Risk (Lewis), 48
Toxic Terror (Whelan), 88
Trends
  beliefs related to deteriorating, 12–16
  determining, 59
  distinction between long- and short-run, 35–36
  long-run trends as real trends, 8
  people's perception of, 124–25
  in perception about pollution, 14
Truth Lobby, proposed, 50–53

Urbanization and farmland, 17–25

Variability concept, 60–61
Vogt, William, 55

Wallis, W. Allen, 83–84

Waste disposal
problem of solid, 9–10, 41
public perceptions of problems, 10–11
Weiss, Julian, 27
Wertham, Frederic, 67
Whelan, Elizabeth, 88

Yergin, Daniel, 28